2nd Radio and Antenna Days of the Indian Ocean

(RADIO 2014)

IOP Conference Series: Materials Science and Engineering Volume 67

Flic-en-Flac, Mauritius
7 – 10 April 2014

Editor:

V. Monebhurrun

ISBN: 978-1-63439-878-7
ISSN: 1757-8981

Printed from e-media with permission by:

Curran Associates, Inc.
57 Morehouse Lane
Red Hook, NY 12571

Some format issues inherent in the e-media version may also appear in this print version.

Printed by Curran Associates, Inc. (2015)

For permission requests, please contact the Institute of Physics
at the address below.

Institute of Physics
Dirac House, Temple Back
Bristol BS1 6BE UK

Phone: 44 1 17 929 7481
Fax: 44 1 17 920 0979

techtracking@iop.org

Additional copies of this publication are available from:

Curran Associates, Inc.
57 Morehouse Lane
Red Hook, NY 12571 USA
Phone: 845-758-0400
Fax: 845-758-2634
Email: curran@proceedings.com
Web: www.proceedings.com

TABLE OF CONTENTS

IOP Conf. Series: Materials Science and Engineering **67** (2014) 011001 doi:10.1088/1757-899X/67/1/011001

Editor's Foreword

It was an honor and a great pleasure for all those involved in its organization to welcome the participants to the "Radio and Antenna Days of the Indian Ocean" (RADIO 2014) international conference that was held from 7th to 10th April 2014 at the Sugar Beach Resort, Wolmar, Flic-en-Flac, Mauritius.

RADIO 2014 is the second of a series of conferences organized in the Indian Ocean region. The aim of the conference is to discuss recent developments, theories and practical applications covering the whole scope of radio-frequency engineering, including radio waves, antennas, propagation, and electromagnetic compatibility. The RADIO international conference emerged following discussions with engineers and scientists from the countries of the Indian Ocean as well as from other parts of the world and a need was felt for the organization of such an event in this region. Following numerous requests, the Island of Mauritius, worldwide known for its white sandy beaches and pleasant tropical atmosphere, was again chosen for the organization of the 2nd RADIO international conference.

The conference was organized by the Radio Society, Mauritius and the Local Organizing Committee consisted of scientists from SUPELEC, France, the University of Mauritius, and the University of Technology, Mauritius. We would like to take the opportunity to thank all people, institutions and companies that made the event such a success. We are grateful to our gold sponsors CST and FEKO as well as URSI for their generous support which enabled us to partially support one PhD student and two scientists to attend the conference. We would also like to thank IEEE-APS and URSI for providing technical co-sponsorship.

More than hundred and thirty abstracts were submitted to the conference. They were peer-reviewed by an international scientific committee and, based on the reviews, either accepted, eventually after revision, or rejected. RADIO 2014 brought together participants from twenty countries spanning five continents: Australia, Botswana, Brazil, Canada, China, Denmark, France, India, Italy, Mauritius, Poland, Reunion Island, Russia, South Africa, South Korea, Spain, Switzerland, The Netherlands, United Kingdom, and USA. The conference featured eleven oral sessions and one poster session on state-of-the-art research themes. Three internationally recognized scientists delivered keynote speeches during the conference. Prizes for the first and second Best Student Papers were awarded during the closing ceremony.

Following the call for the extended contributions for publication as a volume in the IOP Conference Series: Materials Science and Engineering (MSE), both on-line and in print, we received thirty-two full papers. All submitted contributions were then peer-reviewed, revised whenever necessary, and accepted or rejected based on the recommendations of the reviewers of the editorial board. At the end of the procedure, twenty-five of them have been accepted for publication in this volume.

Editor
Vikass Monebhurrun

IOP Conf. Series: Materials Science and Engineering **67** (2014) 011002 doi:10.1088/1757-899X/67/1/011002

Editorial Board Members

Modelling of broadband 1-18 GHz antipodal dual-polarized Vivaldi radiators for antenna array

Y V Yukhanov[1], A I Semenikhin[1], D V Semenikhina[1] and T Y Privalova[1]

[1]Southern Federal University, Taganrog, Russia

E-mail: anilsem@mail.ru

Abstract. Four models of 1 – 18 GHz antipodal Vivaldi radiators with dual orthogonal polarizations as part of the infinite antenna array (AA) are considered. The models have antipodal location of printed conductors on two sides of a substrate in tapered slot part of the antenna; elliptic resonators and a symmetry plane passing through a diagonal of Floquet's unit cell. For VSWR lowering and improvement of isolation of orthogonal channels at resonance frequencies the models can contain the rectangular plates or rods of an absorber. The best model of the radiator AA with long absorber rods has VSWR not higher than 2.55 and an isolation of orthogonal channels more than 19 dB.

1. Introduction

A problem of the creation of the video impulse antenna arrays (AA) is associated to the analysis and selection of radiators of ultrashort pulses of electromagnetic (EM) field [1–4]. Also it should be taken into account (and if it is possible to use) the electromagnetic interaction between the radiators both in outer EM field and in the feeding system [2]. The problem becomes complicated if an AA should operate with two orthogonal polarizations [5].

In this report several models of antipodal ultra-wideband Vivaldi radiators with dual orthogonal polarizations as part of the infinite AA are considered. Numerical simulation of radiators was executed using Ansoft HFSS.

2. Model 1 of Vivaldi radiators with metal rods

Model 1 of Floquet's unit cell of antipodal Vivaldi radiator (AVR) with dual-orthogonal linear polarizations consists of two orthogonal radiators (the left part of figure 1) and has two input connectors with fragments of the coaxial cables. Two Rogers Ultralam 2000™ substrates ($\varepsilon = 2.5$, $\mathrm{tg}\delta = 0.0019$) have thicknesses of 0.5 mm and lengths of 95 mm. The radiator has antipodal location of printed conductors on two sides of a substrate in tapered slot part of the antenna and elliptic resonators. The resonator profile is part of an ellipse with ellipticity of $ell_{1,2} = 2r_{1,2}/(d-w)$ (in this model $r_1 = r_2$; $ell_{1,2} = 0.75 - 1.3$). The simulation considers the thicknesses of printed conductors of 35 μm.

The right part of figure 1 shows compact package of two orthogonally located Vivaldi's radiators in one Floquet's unit cell. Radiator design has a plane of symmetry passing through a diagonal of the Floquet's unit cell. Thus the conditions of the excitation of two orthogonal channels (Port 1, Port 2) are identical, it is essentially important for alignment of phase characteristics of channels [5]. The model includes the shield for mounting of radiators.

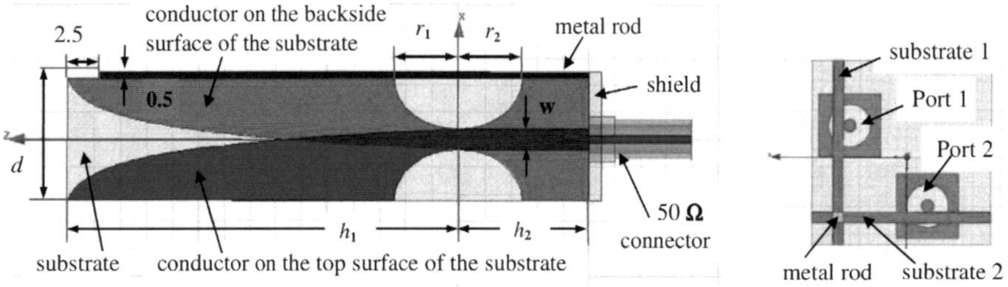

Figure 1. Model 1 geometry of AVR with linear polarization (left) and unit cell cross section of AVR with dual-orthogonal linear polarizations (right). The image is not to scale, but is representative of the configuration.

On the long edges of substrates the metal rods connecting among themselves adjacent radiators take place. They are intended for elimination of the anomalies of VSWR at resonance frequencies [2]. Absorbing boundary conditions were set on rather remote face surface of the radiator Floquet's cell for increase of decision accuracy at low frequencies.

The form and the sizes of radiators, the location of a resonator, the form and the resonator sizes, the lengths of metal rods were variable parameters. Figure 1 and table 1 show the dimensions for model 1 of the AVR unit cell resulting from the optimization.

Table 1. Final radiator parameters design.

Parameter	d	h_1	h_2	r_1	r_2	w
Dimension (mm)	9.4	80	15	2.96 – 5.14	2.96 – 5.14	1.5

The main characteristics of radiator model 1 as a part of the infinite AA in a band of 1 – 20 GHz are given in figure 2. Calculation precision after 7th steps, when the construction was divided on 237539 tetrahedra, was 0.008.

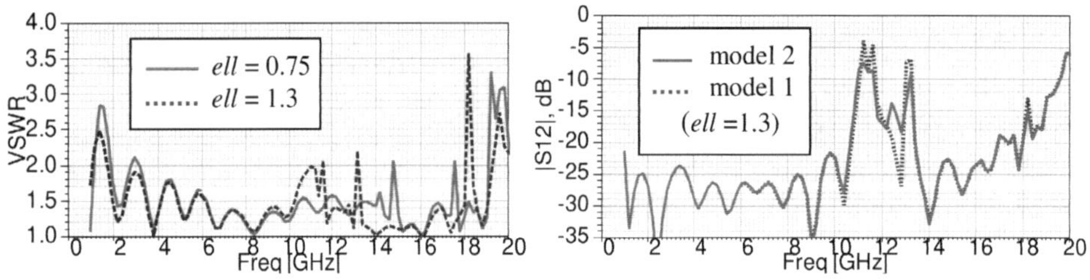

Figure 2. VSWR for models 1 (left) and isolation of orthogonal channels of a single AVR as part of the infinite AA for models 1 and 2 (right).

VSWR on inputs of a radiator (the left part of figure 2) has some maxima caused by resonances of a field. The field intensity in the region of the resonator of the active channel of a radiator sharply increases at several resonance frequencies. The resonator of the passive channel is intensively excited (figure 3) and the isolation of orthogonal channels |S12| sharply decreases to a minus of 4 dB (blue line on the right part of figure 2). In the passive channel two waves running from the resonator (as from a source) in different directions are formed. At resonances the intensity of the E-field can increase on the edges of two resonators (see the left part of figure 3) and on the edges of conductors (the right part of figure 3). Though the efficiency of a radiator isn't worse than 80 %; its application is limited by the isolation of orthogonal channels.

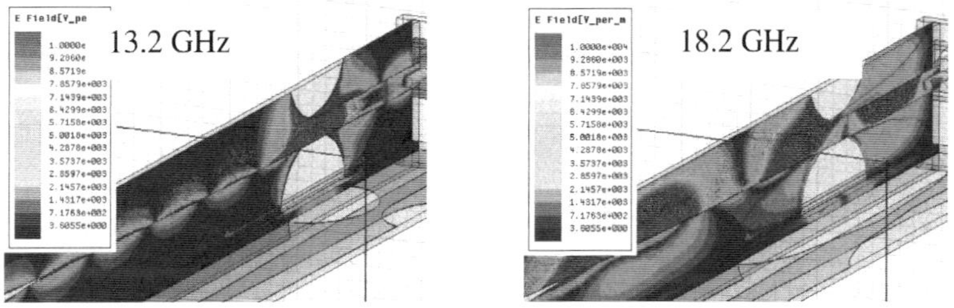

Figure 3. E-field on the conductors of the passive channel at resonance frequencies.

3. Model 2 of the radiator with the absorber plates

In the second model of a Vivaldi radiator (figure 4) in addition to metal rods the rectangular plates of absorber (0.2 mm thick) in the middle of two resonators are placed for blanking of high-Q resonances. The simulation showed that VSWR doesn't exceed 2.5 in operating band of 1 – 18 GHz, but absorber plates eliminate only some anomalies of VSWR (figure 5). The absorber plates improve the isolation of channels (on 4 dB; red line in the right part of figure 2), but the efficiency is slightly reduced (in comparison with model 1). We will mark that the introduction in resonators of similar metal filtering plates (instead of the absorber plates) doesn't remove the resonances.

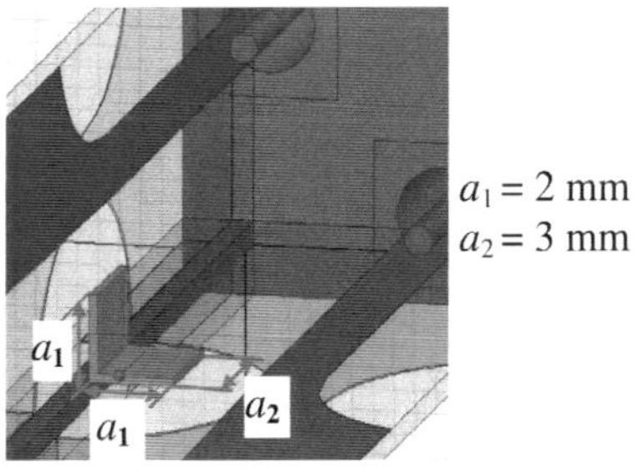

Figure 4. A fragment of model 2 with rectangular plates of an absorber.

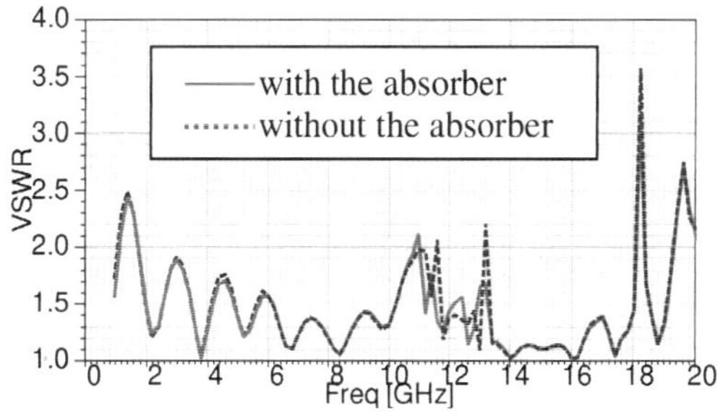

Figure 5. VSWR.

4. Model 3 of the radiator with the short absorber rods

Model 3 differs from the previous by a form of a contour of the resonator and a placement of an absorber. Front, also elliptic, longer part of the resonator with other ellipticity of the contour $ell_1 = 1.5 - 4.5$ is added. The contour of back part of the resonator (near to the input connectors) is still described by ellipticity $ell_2 = 0.75 - 1.5$. The absorber rods are located from the resonator until the end of the radiator (instead of metal rods). Remaining geometrical parameters of the radiator are as in the previous model.

Existence of the absorber rods leads to VSWR lowering almost in all range of frequencies (the left part of figure 6). The isolation between channels increases and it isn't worse than 12 dB (the right part of figure 6). However, the radiator efficiency considerably decreases because of the big area of the absorber.

5. Model 4 of the radiator with the long absorber rods

In model 4 the absorber rods have the cross section of 1.25 mm×1.25 mm, length of 92.5 mm. They pass from shield through the resonator prior to the beginning of the substrate. The ellipticity of a profile of the resonator, as in model 2, varies within the limits $ell_1 = ell_2 = 0.75 - 1.5$.

Model 4 possesses the best characteristics in case $ell_1 = ell_2 = 0.75$ for the cross size of the Floquet's unit cell of $d = 9.2$ mm; VSWR on two inputs of the radiator isn't higher than 2.55 (the left part of figure 6). The maximum value of the realized gain of one radiator increases from a minus of 20 dB to plus of 2 dB in a band of 1 – 18 GHz (the left part of figure 7). The polarizing isolation of two orthogonal channels of model 4 makes more than 19 dB (the right part of figure 6). Phase-frequency characteristics of two channels of this model coincide with graphic precision. A disadvantage of model 4, as well as model 3, is the low efficiency at the upper frequency range (the right part of figure 7). However, it isn't worse than for model 3 (the right part of figure 7).

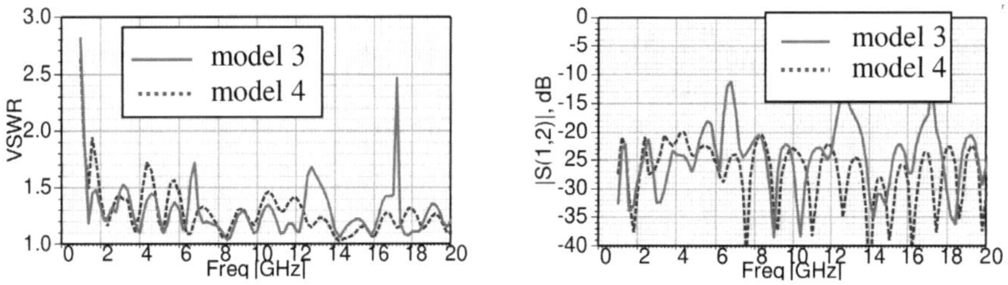

Figure 6. VSWR (left) and isolation of orthogonal channels (right) of a single AVR as part of the infinite AA for models 3 and 4.

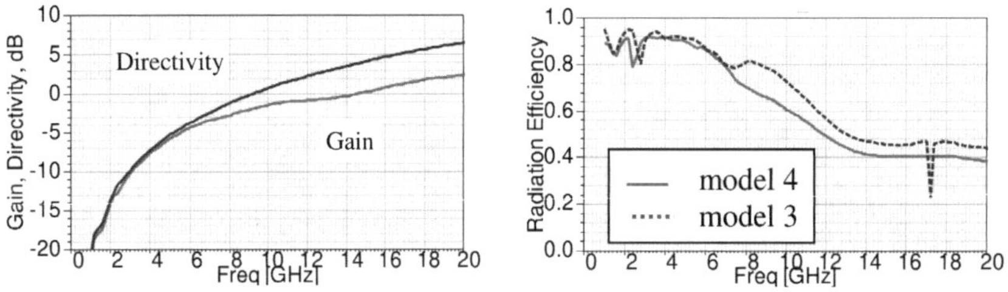

Figure 7. Gain, directivity for model 4 (left) and efficiency (right) of single radiator as part of the infinite AA for models 3 and 4.

6. Conclusion

Four models of antipodal Vivaldi radiators from 1 GHz to 18 GHz with dual orthogonal polarizations as part of the infinite antenna array were considered. At the model with metal rods very weak isolation of orthogonal channels is observed. The model of the radiator with long absorber rods has the best characteristics of VSWR not higher than 2.55 and the isolation of orthogonal channels more than 19 dB, but has the worst efficiency. The model with the rectangular plates of the absorber in the resonator can be a compromise in operating band of 1 – 18 GHz.

7. Acknowledgement

This work was performed as project part of the state task No. 8.2461.2014/K of the Ministry of Education of Russia.

References

[1] Baum C E and Farr E G 139 *Impulse radiating antennas* in Bertoni H 1993 *Ultra-wideband, short-pulse electromagnetics* (Plenum Press)
[2] Shin J and Schaubert D H 1999 *IEEE Trans. on Antennas and Propag.* **47** 879
[3] Schantz H 2005 *The Art and the Science of Ultrawideband Antennas* (Artech House, Inc., USA)
[4] Qing X, Chen Z N and Chia M *Hindawi Publishing Corporation. Int. J. of Antennas and Propag.* Vol 2008 Article ID 267197
[5] Yukhanov Y V, Privalova T Y, Semenikhin A I and Semenikhina D V 2012 *15th Int. Symp. on Antenna Techn. and Appl. Electromag. (ANTEM)* (Toulouse/ France)

Synthesis of quasi-stationary inhomogeneous impedance body of arbitrary shape

Y V Yukhanov [1] and T Y Privalova [1]

[1]Southern Federal University, Taganrog, Russia

e-mail: yu_yukhanov@mail.ru

Abstract — Formulas were obtained for the reflection factors of a plane wave from the inhomogeneous impedance plane, which reflects the incident wave in the given direction. The problem of synthesis of impedance inhomogeneous rotation body was solved for an arbitrary shaped generatrix for the given scattering diagram for co- and cross polarizations. Impedance distributions were obtained in explicit form. Restrictions on the class of implementable scatterplots were formulated.

1. Introduction

An efficient way to control scattering characteristics of an object is to use a wide class of electrodynamics structures which physical characteristics are described by impedance boundary conditions [1], for example, the concentrated or distributed impedance loadings [2], [3]. The design of three-dimensional impedance bodies of arbitrary shape with large electrical sizes requires solving of inverse electrodynamic problems with approximate methods. To do so, we need to know the coefficients of wave reflection from inhomogeneous nonstationary impedance plane.

At the time, the paper authors have obtained reflection coefficients for the following cases of inhomogeneous impedance plane:

- the stationary impedance reflects the wave in nonspecular direction which lies in the incidence plane of a wave at the given polarization [4], [5];
- the stationary impedance reflects the wave in a direction at the given polarization [6];
- the quasi-stationary impedance reflects the wave in a direction at the given polarization with the specified frequency [7].

2. Problem statement

An impedance reflector in the form of rotation body (figure 1) is located in a free isotropic space. The reflector is limited by a piecewise-differentiable surface S with the meridian section $\rho(\theta)$ contour. It's excited by an electromagnetic wave with ω_i frequency coming from an arbitrary direction. The impedance on the surface can be described with Leontovich impedance boundary conditions

$$[\mathbf{n},\mathbf{E}] = -\hat{Z}[\mathbf{n}[\mathbf{n},\mathbf{H}]]. \qquad (1)$$

Let's find the distribution of the passive surface impedance Z (*ReZ*≥*0*) that provides the object S with required scattering diagram $F(\theta,\varphi)$ with frequency ω_s, such as $(\Delta\omega = \omega_s - \omega_i << \omega_i)$.

3. Scattered Field

Let's implement the impedance in a form of frequent orthogonal lattice impedance strips $Z_E(x,y,t)$, $Z_M(x,y,t)$ (figure 2). In this case

$$E_{\mathrm{u}}=Z_E H_{\mathrm{v}};\quad E_{\mathrm{v}}=-Z_M H_{\mathrm{u}}. \tag{2}$$

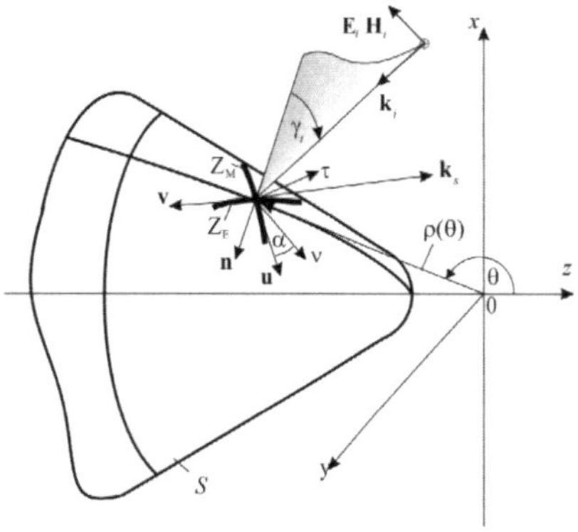

Figure 1. Problem statement.

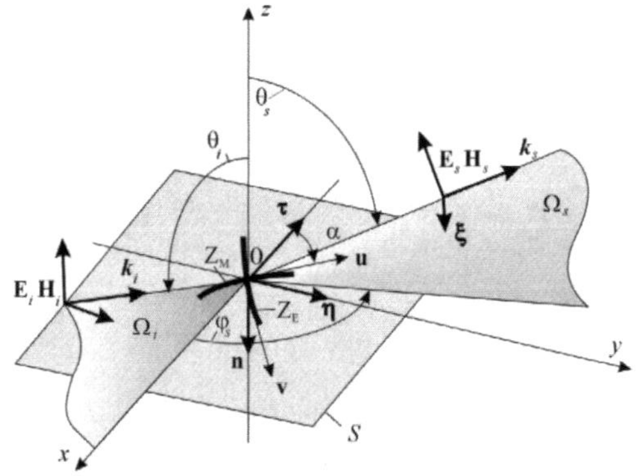

Figure 2. Problem statement.

The orientation of the bands $Z_E(x,y,t)$, $Z_M(x,y,t)$ in the S plane is defined by $\mathbf{u}$ and $\mathbf{v}$ unit vectors ($\mathbf{v}=[\mathbf{n},\mathbf{u}]$, $\cos\alpha=(\mathbf{u},\boldsymbol{\tau})$; $\mathbf{n}$ - normal to S). $\mathbf{u}=\boldsymbol{\tau}\cos\alpha+\boldsymbol{\eta}\sin\alpha$; $\mathbf{v}=-\boldsymbol{\tau}\sin\alpha+\boldsymbol{\eta}\cos\alpha$. Angle $\alpha(x,y,t)$ is a function of time and coordinates on the S plane.

We can use the physical optics approximation to solve the problem. To do this, we need to use the reflection factors of a nonstationary inhomogeneous impedance plane which reflects an incident wave into required direction [7]. It's electric and magnetic fields are orthogonal to the plane of incidence Ω^i and reflection Ω^s (figure 2) and can be described as:

$$E_\xi^s=P_{11}E_\eta^i+P_{12}H_\eta^i;\quad H_\xi^s=P_{21}E_\eta^i+P_{22}H_\eta^i. \tag{3}$$

The scattered field in the far zone can be described by

$$E_\theta^s=\frac{1}{4\pi Rc}\int_S\frac{\partial}{\partial t'}\left\{\sin\varphi\left[J_\eta^e(q,t')\cos\theta-J_\tau^m(q,t')\right]+\cos\varphi\left[J_\tau^e(q,t')\cos\theta-J_\eta^m(q,t')\right]\right\}dS_q;$$

$$E_\varphi^s=\frac{1}{4\pi Rc}\int_S\frac{\partial}{\partial t'}\left\{\sin\varphi\left[J_\tau^e(q,t')+J_\eta^m(q,t')\cos\theta\right]-\cos\varphi\left[J_\eta^e(q,t')+J_\tau^m(q,t')\cos\theta\right]\right\}dS_q, \tag{4}$$

where $t'=t-R/c$; R – distance from the integration point to the observation point; c – light velocity.

Surface currents $\mathbf{J^e}=\boldsymbol{\tau}J_\tau^e+\boldsymbol{\eta}J_\eta^e$ and $\mathbf{J^m}=\boldsymbol{\tau}J_\tau^m+\boldsymbol{\eta}J_\eta^m$ can be determined using the reflection coefficients (3), obtained in the work [7].

4. Impedance synthesis

The required impedance distribution can be found from the boundary conditions (2):

$$Z_E=\frac{(\mathbf{l^i},\mathbf{u})e^{i\Phi_i}+(\mathbf{l^s},\mathbf{u})H_0^s e^{i\Phi_s}}{(\mathbf{h^i},\mathbf{v})e^{i\Phi_i}+(\mathbf{h^s},\mathbf{v})H_0^s e^{i\Phi_s}};\qquad Z_M=-\frac{(\mathbf{l^i},\mathbf{v})e^{i\Phi_i}+(\mathbf{l^s},\mathbf{v})H_0^s e^{i\Phi_s}}{(\mathbf{h^i},\mathbf{u})e^{i\Phi_i}+(\mathbf{h^s},\mathbf{u})H_0^s e^{i\Phi_s}}, \tag{5}$$

where the incident and the scattered fields are: $\mathbf{H^i}=\mathbf{h^i}e^{i\Phi_i}$; $\mathbf{E^i}=\mathbf{l^i}e^{i\Phi_i}$; $\mathbf{H^s}=\mathbf{h^s}He^{i\Phi_s}$; $\mathbf{E^s}=\mathbf{l^s}He^{i\Phi_s}$; $\Phi_{i,s}=-k_{i,s}(\mathbf{k}^{i,s},\mathbf{R})$; $\mathbf{R}$ - radius-vector; in indexes of exponents $i=\sqrt{-1}$.

The conditions of absolutely reactive impedance implementation can be found from (5):

$$\mathrm{Re}(Z_E)=0;\ \mathrm{Re}(Z_M)=0.$$

They define the restrictions on the class of realized scattering diagrams:

$$\sqrt{(l_\xi^s)^2+(h_\xi^s)^2}=1;\ h_\xi^s=\frac{\cos\gamma_i\cos(\alpha+\beta)}{\sqrt{\sin^2(\alpha+\beta)+\cos^2\gamma_i\cos^2(\alpha+\beta)}};\ H=\frac{\sqrt{\cos^2\gamma_i\cos^2(\alpha+\beta)+\sin^2(\alpha+\beta)}}{\cos\gamma_i}. \quad (6)$$

Absolutely reactive impedance, according to (6), performs full reflection of the waves only in a cone $\theta_i=\theta_s$ with amplitude H and polarization h_ξ^s. If we also set the polarization of the reflected wave, for example $l_\xi^s=0; h_\xi^s=1$, then (6) implies that a complete reflection is only possible in the plane of incidence of the wave ($\alpha+\beta=\pm\pi$). This means, as shown by studies [4] - [7], that, in addition to the beam in the given direction at the given frequency and polarization, there is a specular beam at the co-polarization with initial frequency ω_i.

So the system of parallel strips ($\alpha=0;\ Z_M=0;\ Z_{11}=0;\ Z_{22}=Z_E;\ Z_{12}=Z_{21}=0$), reflecting the wave in one plane:

$$Z_{22}=i\cos\theta_s tg\{0.5[k_s x((\omega_i/\omega_s)\sin\theta_i+\cos\varphi_s\sin\theta_s)-\Delta\omega t]\}, \quad (7)$$

creates the scattered field with the following normalized scattering diagram:

$$F_\theta(\theta,\varphi)=F_\theta^i(\theta,\varphi)+F_\theta^s(\theta,\varphi) \quad (8)$$

where $F_\theta^i(\theta,\varphi)=\cos\varphi(\cos\theta_i+\cos\theta_s)\dfrac{\cos\theta-\cos\theta_s}{4\cos\theta_s}\dfrac{\sin[k_i b\sin\varphi\sin\theta]}{k_i b\sin\varphi\sin\theta}\dfrac{\sin[k_i a(\cos\varphi\sin\theta+\sin\theta_i)]}{k_i a(\cos\varphi\sin\theta+\sin\theta_i)}$;

$$F_\theta^s(\theta,\varphi)=\cos\varphi(\cos\theta_i+\cos\theta_s)\frac{\cos\theta+\cos\theta_s}{4\cos\theta_s}\frac{\sin[k_s b\sin\varphi\sin\theta]}{k_s b\sin\varphi\sin\theta}\frac{\sin[k_s a(\cos\varphi\sin\theta-\cos\varphi_s\sin\theta_s)]}{k_s a(\cos\varphi\sin\theta-\cos\varphi_s\sin\theta_s)}.$$

The main lobe of the diagram $F_\theta^i(\theta,\varphi)$ ($\cos\varphi\sin\theta+\sin\theta_i=0$) for ω_i frequency of the incident wave is oriented in specular direction ($\varphi=\pi;\ \sin\theta-\sin\theta_i=0;\ \theta=\theta_i$). For $\theta_i=\theta_s$ it disappears, i.e. there is a complete reflection of the wave $F_\theta(\theta,\varphi)=F_\theta^s(\theta,\varphi)$ for a given frequency ω_s in the direction $\cos\varphi\sin\theta-\cos\varphi_s\sin\theta_s=0$.

If the quasi-stationary impedance $Z_{22}(t)$ depends on time only, then when a wave incidents from a direction different from the normal ($\theta_i\neq 0$), the scattered field at the frequency $\omega_s=\omega_i+\Delta\omega$ has a main lobe with a maximum in nonspecular direction $\theta=\theta_s=\arcsin(\omega_i\sin\theta_i/\omega_s)$.

Figure 3 shows the RP $F_\theta(\theta,\varphi)=F_\theta^i(\theta,\varphi)+F_\theta^s(\theta,\varphi)$ of an antenna ($\varphi=0$) with cone-shaped impedance reflector $\rho(\theta)=-h/\sin(\theta+\beta)$; with vertex angle of $2\beta=120^0$ and base radius of $a=100mm$; the distance from the cone vertex to the rectilinear irradiator is $h=100mm$.

Impedance simulates a parabolic reflector, which aperture coincides with the cone base (the maximum radiation is directed along the axis). The curve 1 is the RP of an equivalent reflector antenna, the curve 2 is the radiation pattern (RP) of the impedance antenna $F_\theta^s(\theta,\varphi)$ at the given frequency $\omega_s=\omega_i+\Delta\omega$ and the curve 3 is the RP of the impedance antenna $F_\theta^i(\theta,\varphi)$ at the frequency of the irradiator ω_i. In this case $F_\theta^i(\theta,\varphi)$ coincides with the pattern of Kirchhoff's "black" body model scattered field. $F_\theta^s(\theta,\varphi)$ in the area of main and the first side lobes is almost identical to the RP of the equivalent parabolic antenna.

The calculations above are given for the physical optics approximation. For the verification of carried out research the figure 4 shows the scattering diagrams of impedance reflector in the form of a disk with the radius of $a=100mm$, reflecting the incident wave $\theta_i=0$ ($\lambda=30mm$) at $\theta_s=45°$.

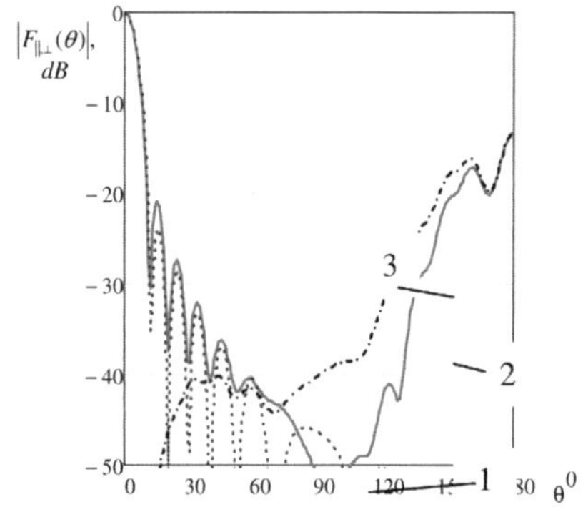

Figure 3. Radiation patterns.

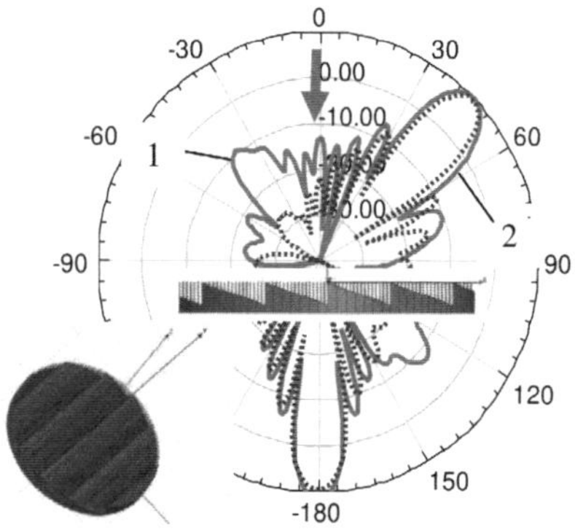

Figure 4. Bistatic RCS diagrams.

Stationary impedance (7) ($\Delta\omega=0$; $Z_{22}=i\cos\theta_s tg\{0.5k_i x(\sin\theta_i+\cos\varphi_s\sin\theta_s)\}$) implemented by a ribbed structure with a period of 2.5 mm and the width of the grooves of 2 mm. Calculations were performed with Ansoft HFSS. The curve 1 describes the scattering diagram of the impedance disk, the curve 2 describes the scattering diagram of the equivalent perfectly conducting disk, turned away from the direction of incidence of the wave at an angle $\theta_s/2=22.5°$. Precise calculations using HFSS (after 14th steps, when the construction was divided on 731945 tetrahedral, the archived precision was 0.005) demonstrated good coincidence of given (curve 2) and implemented (curve 1) diagrams.

5. Conclusion

The problems of synthesis of isotropic and anisotropic, stationary and quasi-stationary impedance reflectors and antennas with the reflector of arbitrary shape are solved in an explicit form. A class of implemented scattering diagrams is found. The numerical results confirm the validity of the derived formulas. These formulas (5) - (7) allow to solve the synthesis problem with high accuracy.

6. Acknowledgement

This work was performed as project part of the state task No. 8.2461.2014/K of the Ministry of Education of Russia.

References

[1] Miller M A and Talanov V I 1961 *Radiophysic* **4** 795
[2] Lawrie R E 1966 *IEEE Trans. on Antennas and Propag.* **3** 1
[3] Short I and Chen K M 1969 *IEEE Trans. on Antennas and Propag.* **6** 34
[4] Yukhanov Y V 2000 *Radio engineering and electronics* **45** 404
[5] Yukhanov Y V, Yukhanov A Y and Privalova T Y 2013 *Proceeding of the 2013 International Conference on Electromagnetics in Advanced Applications* doi: 10.1109/ICEAA.2013.6632178
[6] Yukhanov Y V and Privalova T Y 2013 *Asia-Pacific Microwave Conference Proceedings* doi: 10.1109/APMC.2013.6694871
[7] Yukhanov Y V and Privalova T Y 2013 *Antenny* **10** 11.

IOP Conf. Series: Materials Science and Engineering **67** (2014) 012003 doi:10.1088/1757-899X/67/1/012003

Comparative Study on Synthesizing Reconfigurable Time-Modulated Linear Arrays using Differential Evolution, Artificial Bee Colony and Particle Swarm Optimization

S. K. Mandal, Harshavardhan Singh, G. K. Mahanti, Rowdra Ghatak

Dept. of Electronics and Communication Engg.,
National Institute of Technology Durgapur, W. B., INDIA
Email: skmandal2006@gmail.com

Abstract. This paper presents a new technique based on optimization tools to design phase only, digitally controlled, reconfigurable antenna arrays through time modulation. In the proposed approach, the on-time durations of the time-modulated elements and the static amplitudes of the array elements are perturbed in such a way that the same on-time sequence and discrete values of static amplitudes for four bit digital attenuators produces either a pencil or a flat-top beam pattern, depending on the suitable discrete phase distributions of five bit digital phase shifters. In order to illustrate the technique, three optimization tools: differential evolution (DE), artificial bee colony (ABC), and particle swarm optimization (PSO) are employed and their performances are compared. The numerical results for a 20-element linear array are presented.

1. Introduction

Multiple radiation patterns from a single antenna array find applications in the field of satellite and military communication, surveillance and radar engineering. Depending on the applications, it is required to synthesize the antenna patterns with some stringent specifications such as low side lobe level (SLL), low ripple (R), sharp transition (T) for flat-top (sector) pattern or narrow first null beam-width (FNBW) for pencil beam pattern. Syntheses of such patterns using conventional antenna array (CAA) synthesis methods need relatively high value of dynamic range ratio (DRR) [1] which results in different sources of errors such as systematic and mechanical errors those degrade the array performance. On the other hand, time-modulation technique is attractive to synthesizing low / ultralow SLL pattern with low DRR [2 - 4] or in uniform amplitude array [5 - 6].

Different population based stochastic techniques such as differential evolution (DE) [2-4, 6]; particle swarm optimization (PSO) [7] and artificial bee colony (ABC) [5, 8] play an important role to reducing/controlling the power losses due to sideband radiations that result from the periodical commutations of the antenna elements [9]. To the best of author's knowledge, comparative study in synthesizing reconfigurable antenna arrays through time modulation using different optimization tools has not been reported. The effectiveness of the technique as proposed in [4] has been used in synthesizing low side lobe pencil beam patterns only. In this paper, this technique is used to synthesize a digitally controlled dual beam (flat-top and pencil) pattern with phase only difference and the performances of three efficient optimization tools such as DE, ABC and PSO; in solving such problem have been studied.

Table 1: Desired and obtained values of the design parameters of the synthesized patterns.

	SLL (dB)				Ripple (dB)				FNBW (deg)			
	Desired	Obtained			Desired	Obtained			Desired	Obtained		
Flat-top	-25.0	-24.78 DE	-22.22 ABC	-24.31 PSO	0.5	0.67 DE	0.48 ABC	0.53 PSO	---	---		
pencil	-25.0	-25.0	25.69	-25.0	---	---			23	21.32 DE	18.09 ABC	21.32 PSO

2. Theory

Let us consider a broadside time modulated linear array antenna (TMLAA) of 2N isotropic elements that are symmetrically placed along the x-axis with inter element spacing of *d* such that array center coincides with the origin of reference coordinate axis. Let us further assume that, excitation amplitudes A_p; phases α_p and normalized on-time sequences τ_p $\forall p \in [1, N]$ of radiating elements are also symmetrical with respect to the array center. Due to the periodical excitation of the array elements, the array factor expression at *k*-th harmonics can be obtained as, in (1), [2, 9].

$$AF_k(u,t) = 2e^{j2\pi(f_0+kf_m)t}\sum_{p=1}^{N}\left\{\tau_p A_p e^{j\alpha_p}\sin c(k\pi\tau_p)e^{-j\pi k\tau_p}\right\}\cos\left[\frac{2(p-1)\pi d}{\lambda}u\right] \tag{1}$$

Where $f_m = 1/T_m$ is the modulation frequency with T_m being the modulation period; f_0 is the frequency of the operating signal with λ being the wavelength; $u = \cos\theta$ when θ is the observation angle measured from the array axis.

3. Optimization Strategy

In order to reducing the sideband level (SBL), the optimization process is divided in two steps as detail is explained in [4]. In the first step, the dynamic excitations $E_p = A_p\tau_p$ are perturbed in the continuous search space (0.05, 1). α_p is distributed in the discrete search space (-180, 180) by equally dividing 32 steps for the five bit digital phase shifters. The fitness function for this step is defined as

$$\Omega^{(1)}(E_p,\alpha_p) = \sum_{\delta=1,2}\left\{\sum_{m=1}^{2} W_m^{(\delta)}\left|S_{m,o}^{(\delta)} - S_{m,d}^{(\delta)}\right| H(S_{m,o}^{(\delta)} - S_{m,d}^{(\delta)})\right\} \tag{2}$$

where, $\delta = 1, 2$ correspond to the flat-top and pencil beam pattern respectively; m = 1, 2 are the design parameters and '*S*' denotes the design specifications of the patterns as given in Table 1. The subscripts '*d*' and '*o*' in '*S*' indicate the desired and obtained valued of the parameters. W_m*'s* are the weighting factor and '*H*' is the Heaviside step function. In the second step, discrete search space of A_p i.e., (0.25, 1) is equally divided into 16 steps for the four bit digital attenuators. Continuous search space of τ_p is taken as (0.1, 1). Then, as in [4], the following cost function is used to reduce SBL

$$\Omega^{(2)}(A_p,\tau_p) = \sum_{\delta=1,2} SBL_{\max}^{(\delta)} \tag{3}$$

with SBL_{max} as the maximum sideband level among the first three sidebands.

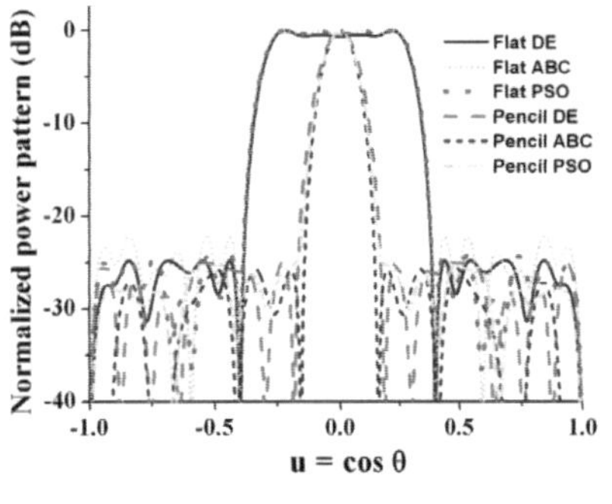

Fig.1: Dual beam patterns using DE, ABC and PSO.

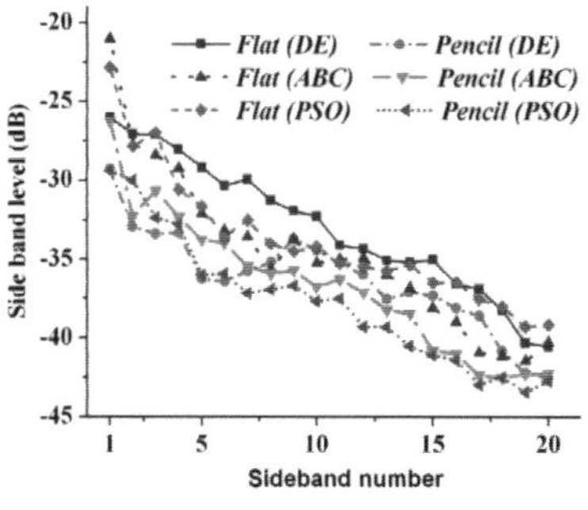

Fig.2: Side band levels for the first 20 sidebands.

Table 2: Excitation amplitudes, on-time durations and phases of elements for different algorithms

Element Numbers → / Excitations ↓		2 & 19	3 & 18	4 & 17	5 & 16	6&15	7&14	8&13	9&12	10&11
A_k (DE)	1	0.2500	0.250	0.2969	0.4375	0.6719	0.7656	0.7187	0.8125	1
A_k (ABC)	0.250	0.250	0.2969	0.3438	0.3906	0.625	0.7188	0.7656	0.6718	0.7656
A_k (PSO)	0.2500	0.2500	0.3438	0.3438	0.4370	0.7187	0.8593	0.8125	0.8125	0.9063
T_k (DE)	0.0456	0.6238	0.8501	0.8498	0.9828	0.9562	0.9825	0.9883	0.9911	1
T_k (ABC)	0.4823	0.6074	0.95704	0.8826	0.9811	0.9869	0.9943	0.9745	0.9482	1
T_k (PSO)	0.4178	0.8448	0.8806	0.9375	1	0.9998	0.9609	0.9874	0.9789	0.9947
E_k (DE)	0.0390	0.1333	0.1817	0.2157	0.3677	0.5493	0.6432	0.6074	0.6885	0.8551
E_k (ABC)	0.1548	0.1949	0.3646	0.3893	0.4918	0.7916	0.9172	0.9575	0.81767	0.98266
E_k (PSO)	0.1004	0.2030	0.2909	0.3097	0.4205	0.6907	0.7937	0.7711	0.7645	0.8665
Φ_F(deg) (DE)	90.00	0	-22.50	-67.499	-135.00	-168.750	-180.00	146.250	101.250	78.7499
Φ_F(deg) (ABC)	-157.50	112.50	90	45	-11.250	-45	-67.500	-90	-135	-157.50
Φ_F(deg) (PSO)	-11.250	78.750	101.25	146.250	-157.500	-123.750	-101.250	-78.750	-33.750	-11.250
Φ_P(deg)(DE))	-180.00	90.00	-180.00	168.750	180.00	-180.00	180.00	-180.00	180.00	180.00
Φ_P(deg) (ABC)	-180	-180	180	-180	180	-180	180	180	180	180
Φ_P(deg)(PSO)	-101.25	-135.00	-101.25	-90.00	-90.00	-90.00	-78.75	-90.00	-78.75	-78.75

4. Numerical Results and Discussions

A TMLAA with 2N = 20, d = 0.5λ is considered. To study the performance of the three algorithms, control parameters are set as in [10, 11 and 12]. In the two steps, the same values of common parameters such as number of population 100 and 50; maximum generation 4000 and 100 are set respectively. Also by setting same values of weighting factors i.e., $W_1 = 2$, $W_2 = 6$ for flat-top and $W_1 = 5$, $W_2 = 2$ for the pencil beam; the algorithms are run for 20 trials. A selected pattern is shown in Fig. 1 and the corresponding distributions of E_p, and α_p are given in Table 2. It is worth to mention that in CAAs, the distribution of E_p is considered as the static excitation distribution for which DRR value is high e.g., DRR = 21.09 for DE. In the second step, by redistributing A_p and τ_p, SBL_{max} is reduced to below -20 dB (see Fig. 2) as well as the DRR value to '4' (see Tale 2). The average convergence characteristics for the two steps of the different algorithms are shown Fig. 3 and 4 respectively. It can be seen that performance of DE in terms of quick convergence and minimizing the fitness function values are better with respect to both ABC and PSO. Although, the initial convergence rate of ABC is higher than that of PSO but it quickly converges to local minima.

5. Conclusion

This paper presents a new technique to design reconfigurable time-modulated antenna arrays with digitally controlled static excitation amplitudes and phase distributions. The technique successfully

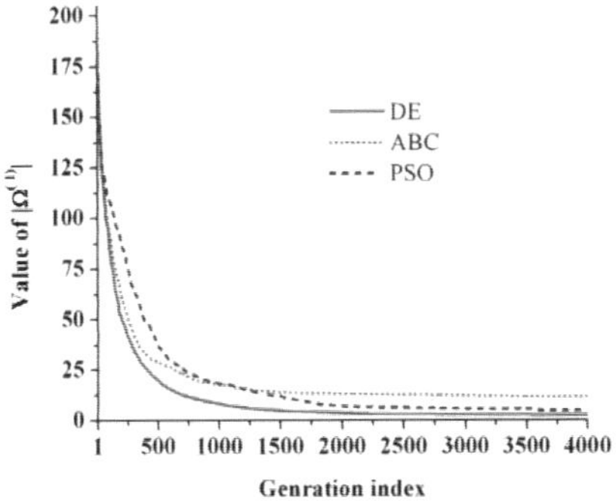

Fig 3: Average convergence characteristics of DE, ABC& PSO for the first step of optimization process.

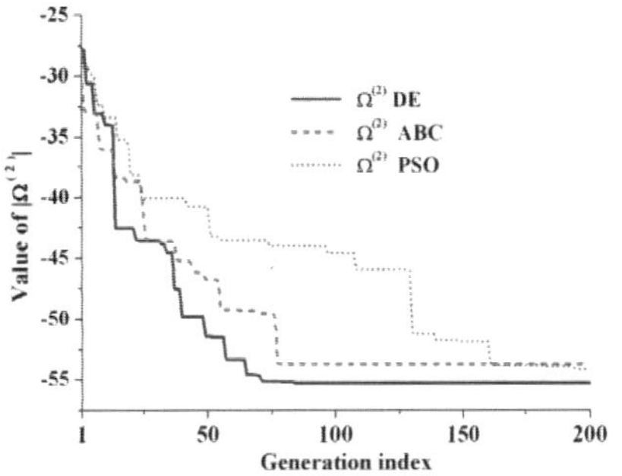

Fig. 4. The average convergence curve of DE, ABC and PSO in the second step of optimization process.

synthesizes the patterns with closed to the desired design specifications. The performances of the three optimization algorithms: DE, ABC and PSO; on solving such problems are studied. Although the performances of the different optimization algorithms are sensitive on the weighting factors of the cost function and for different set of weighting factors final results may be different; in this paper it is observed that DE outperforms over both ABC and PSO.

References

[1] M. Durr, A. Trastoy, and F. Ares, "Multiple-pattern linear antenna arrays with single prefixed amplitude distributions: modified Woodward- Lawson synthesis," *Electron. Lett.,* vol. 36, pp. 1345-1346, 2000.

[2] S. Yang, Y. B. Gan, and P. K. Tan, "A new technique for power-pattern synthesis in time-modulated linear arrays," *IEEE Antennas Wireless Propag. Lett.,* vol. 2, pp. 285-287, 2003.

[3] S. K. Mandal, G. K. Mahanti, and R. Ghatak, "Differential evolution algorithm for optimizing the conflicting parameters in time-modulated linear array antennas," *Progress In Electromagnetics Research, PIER B* vol. 51, pp. 101 - 118, 2013.

[4] S. K. Mandal, G. K. Mahanti, and R. Ghatak, "A single objective approach for suppressing sideband radiations of ultra-low side lobe patterns in time-modulated antenna arrays," *Journal of Electromagn. Waves and Application,* vol. 27, pp. 1767-1775, 2013.

[5] S. K. Mandal, R. Ghatak, and G. K. Mahanti, "Minimization of side lobe level and side band radiation of a uniformly excited time modulated linear antenna array by using artificial bee colony algorithm," in *In the proceedings of IEEE Symposium on Industrial Electronics and Applications (ISIEA 2011)*, 2011, pp. 247-250.

[6] S. K. Mandal, G. K. Mahanti, Rowdra Ghatak, "Design of digitally-controlled multiple-pattern time-modulated antenna arrays with phase-only difference," *Annals of Telecommunications - annales des telecommunications*, *Springer*," *Springer*," Published online: 12 March 2014, DOI: 10.1007/s12243-014-0426-7.

[7] P. Rocca, L. Poli, G. Oliveri, and A. Massa, "Synthesis of time-modulated planar arrays with controlled harmonic radiations," *J. Electromagn. Waves Appl.,* vol. 25, *no.* 5-6, pp. 827–838, 2010.

[8] S. K. Mandal, G. K. Mahanti, Rowdra Ghatak, "Design of a Time-Modulator to Synthesize Different Patterns in Time-Modulated Antenna Arrays," *Journal of Electromagnatic Waves and Application,* Vol. 28, No. 09, 1118-1130.

[9] E. Aksoy and E. Afacan, "Calculation of sideband power radiation in time-modulated arrays with asymmetrically positioned pulses,"*IEEE Antennas and Wireless Propag. Lett.,* vol. 11, pp. 133-136, 2012.

[10] S. Das and P. N. Suganthan, "Differential evolution: A survey of the state–of-the-art," *IEEE Trans. Evol. Comput.,* vol. 15, pp. 4-31, 2011.

[11] B. Akay and D. Karaboga, "Parameter Tuning for the Artificial Bee Colony Algorithm," presented at the ICCCI, Wrocław, Poland, October 5-7, 2009; LNAI 5796.

[12] Y. Shi and R. C. Eberhart, "Parameter selection in particle swarm optimization," in *Poc. Evolutionary Programming VII, EP 98*, Sandiego, California, USA, 1998, pp. 591-600.

Compact concentric ring shaped antenna for ultra wide band applications

Rahul Singha[1], D. Vakula[2] and N V S N Sarma[3]
[1]Research Scholar, National Institute of Technology, Warangal, India
[2]Assistant Professor, National Insititute of Technology, Warangal, India
[3]Professor, National Insititute of Technology, Warangal, India

E-mail: rahul.patch.antenna@gmail.com

Abstract. A novel antenna for compact size, simple structure suitable for low cost fabrication is proposed for UWB application. A compact ring shaped monopole antenna is designed to cover the entire ultra wide bandwidth which has straight forward printed circuit board integration. The dimensions of the antenna are 16mm × 12mm × 0.787mm. More specifically, the impedance matching of the antenna is improved by employment of the tapered microstrip feed line. The measurement and simulation results show that the proposed antenna achieves good impedance bandwidth from 6.5 GHz to 25 GHz which covers the entire UWB. The antenna also has a gain approximately 2.5dBi from 6 GHz to 22 GHz. Compared to the existing UWB antennas, the presented modified structure has the smallest size, the widest bandwidth and better return loss characteristics.

1. Introduction

Ultra-wideband (also known as UWB, ultra-wide band and ultra band) is a radio technology pioneered by robert a. scholtz. The federal communications commission (FCC) and the international telecommunication union radio communication sector (ITU-R) currently define UWB when signal bandwidth exceeds 20% of the centre frequency.UWB technology has been widely used in radar and portable devices (like cell phone and laptop etc.), wireless body area network (WBAN) as well as a wireless personal area network (WPAN) applications. The major challenges in design of UWB antenna are low profile, compact size, good impedance matching and stable omni-directional radiation pattern.

In recent years several UWB antennas have been reported, such as a compact size rectangular patch with centre slot and three via holes which gives a constant gain and stable radiation patterns over the frequency range 2.0 to 12.0 GHz [1]. Lin, S. explained a semi-circle monopole antenna to have a omni-directional characteristic in H-plane [2]. A novel planar triangular monopole antenna (PTMA) is presented by Lin, C.C. which has impedance bandwidth up to about 4:1 [3]. And finally a compact ultra-wideband annular ring antenna is designed with band notch characteristics for wireless local area network (WLAN) and dedicated short-range communication (DSRC) [4]. But few UWB antennas have a structural complexity i.e. the radiators is etched on a substrate and placed vertically 0.5mm above the ground plane [5], which is difficult to integrate in UWB electronics. In the above mentioned

[1] To whom any correspondence should be addressed.

antennas the feeding techniques employed are coplanar waveguide, simple microstrip line and slotted structures.

Day by day, demands of UWB antenna have been increasing gradually; due to improved performances, higher bit rate transmission speeds and there may be need for new and future UWB wireless scheme. In this paper, another promising bandwidth enhancement technique of using a concentric ring shaped monopole antenna is presented. The proposed antenna has improved performance when compared with some recently reported UWB monopole antennas from open literatures [3, 4][6–10].

2. Antenna Design

The printed planar monopole antennas have been widely used in wireless communication devices. They are very popular for volume limited and wideband applications. The topology of the presented antenna composed of a tapered microstrip feed line and a concentric ring shape monopole. The annular ring resonates more than one mode by which the bandwidth of antenna is increase. It is implemented using rogers RT/duroid 5880 board material, which has relative permittivity $\varepsilon_r = 2.2$, height 0.787mm. The top and side views of the presented antenna are shown in figure 1. The antenna overall size is 16mm × 12mm × 0.787mm and to improve impedance matching, it is necessary to transform the impedance of the input feeding microstrip line (50Ω) to the input impedance (100Ω) of the transition.

The original UWB antenna operating at higher frequencies for the simple microstrip feed line is generally limited to around 10 GHz. Here, using tapered microstrip feed line with proper parameters instead of simple feed line, there is an improvement in impedance matching and the impedance bandwidth of the presented antenna can be extended up to 25 GHz. Different kind of feeding techniques are available, but the presented tapered microstrip line configuration offered a simple and very low cost solution for the UWB antenna designs, while also offering good performance values in terms of 50 Ω impedance matching. Using parametric study, the dimension of the tapered microstrip line are optimized, while maintaining the width of the input feed line, the concentric ring monopole, and the characteristics of the utilized substrate.

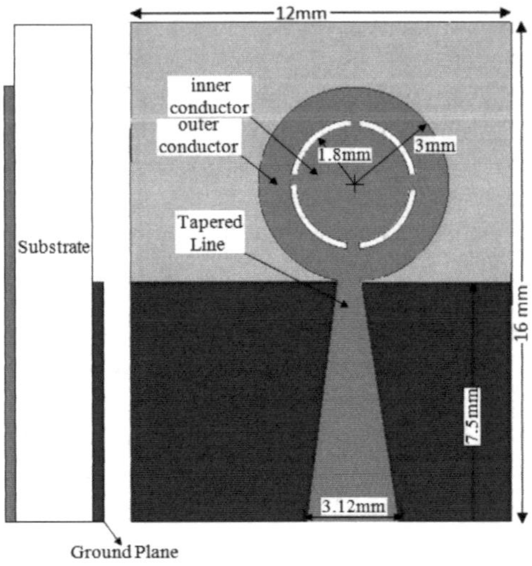

Figure 1. Geometry of the monopole antenna.

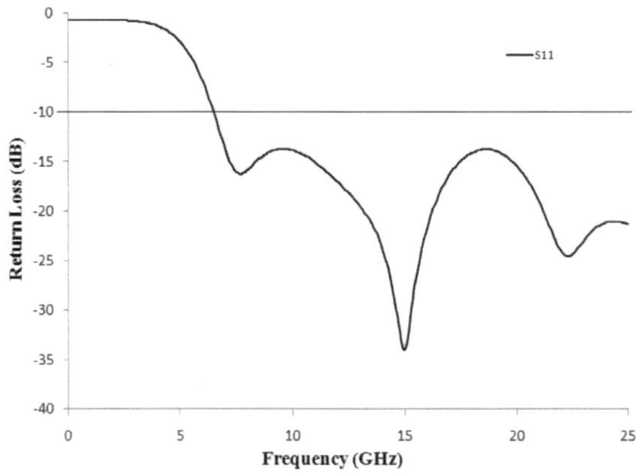

Figure 2. Simulated return loss of the antenna.

3. Result and Discussion

The simulation of the proposed antenna and optimization was done by using the high frequency structure simulator (HFSS) version 15. The simulated return loss of the presented antenna exhibits a wideband performance from 6.5 GHz to beyond 25 GHz for $S_{11} < -10$ dB is shown in figure 2. The photograph of the fabricated original antenna for front and back side is shown in Figure 4 and figure 5 respectively. The return loss of the antenna is measured from 1 to 13.5 GHz as shown in figure 6. The measured result shows good agreement with simulation results. The simulated radiation pattern for the monopole antenna is shown in figure 3 at 15 GHz, 20 GHz. Nearly good omni-directional patterns have been observed in H-plane, and the patterns in the E-plane are close to bidirectional for both 15GHz and 20GHz. The antenna has a gain of 2 to 2.9 dBi from 6 to 22 GHz as shown in figure 7. Better antenna performances for practical applications can be attained as a RT/duroid 5880 substrate is used for fabricating the antenna. Because of its low dissipation factor of RT/duroid 5880 performance is extended to Ku-band and above.

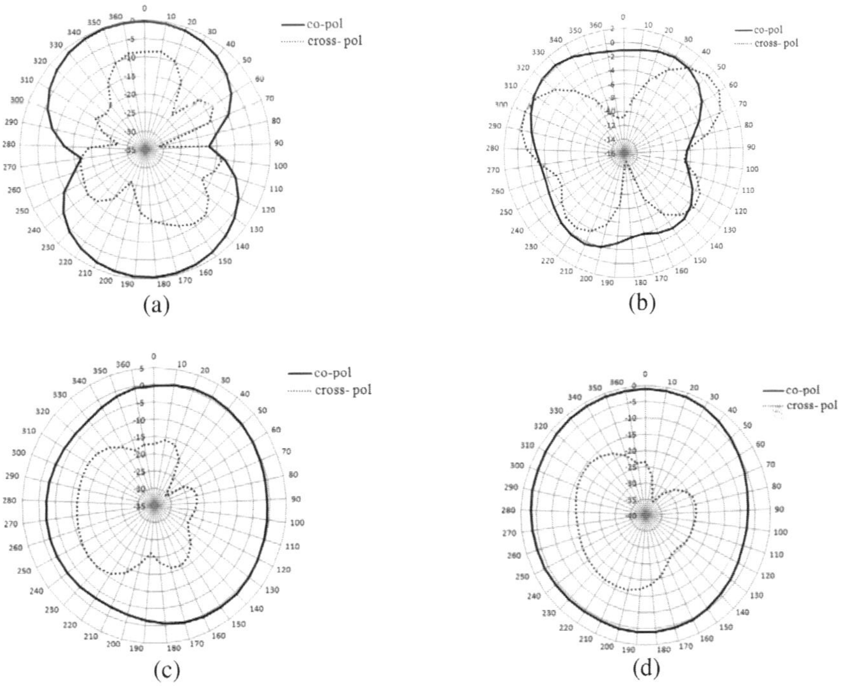

Figure 3. Simulated E-plane radiation patterns of the UWB antenna at (a) 15GHz, (b) 20GHz. And H-plane radiation pattern at (a) 15GHz, (b) 20GHz.

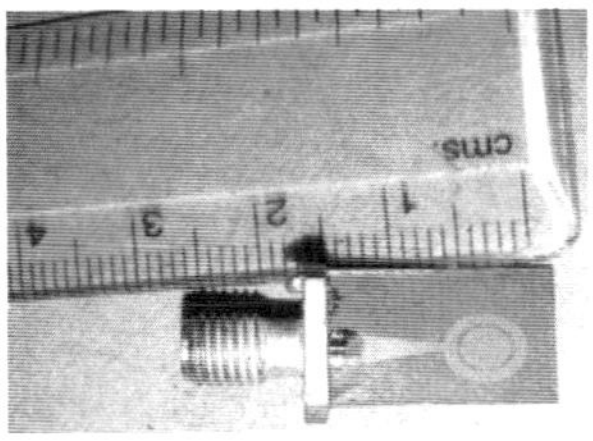

Figure 4. Fabricated front side of the antenna

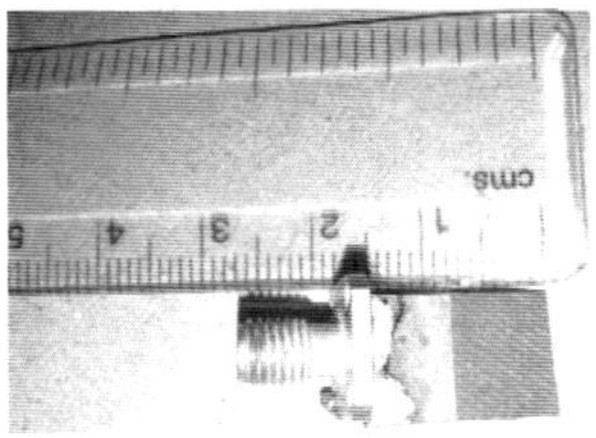

Figure 5. Fabricated back side of the antenna

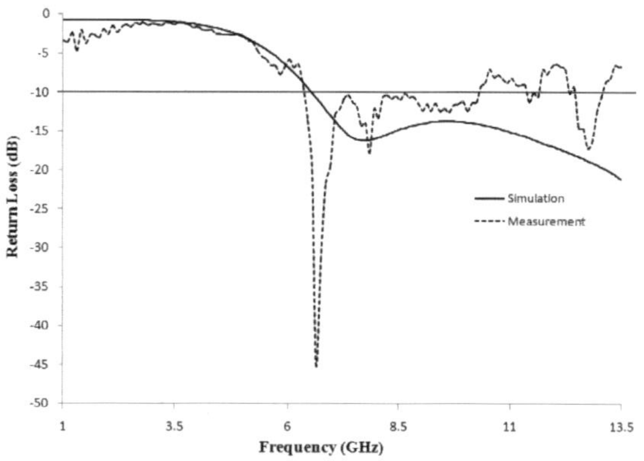

Figure 6. Measured and simulated return loss of the antenna.

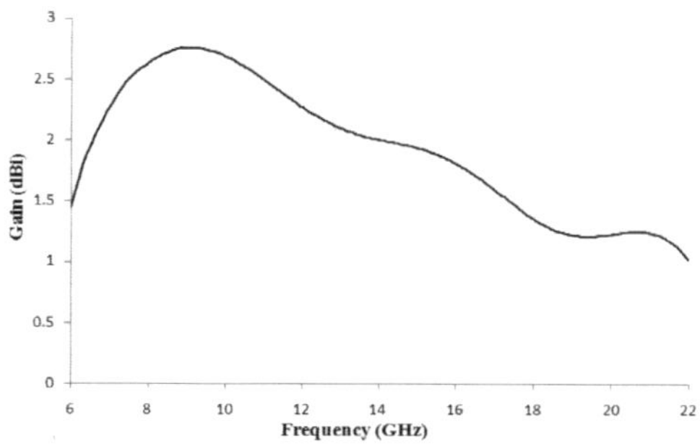

Figure 7. Simulated gain in E-plane.

4. Conclusion

A UWB monopole antenna with very compact size and extremely wide band from 6.5 GHz to beyond 50 GHz is presented in this paper. Concentric ring shape and tapered microstrip feed line are employed for size reduction and band enhancement, respectively. Therefore, the presented antenna has smaller size and wider band compared with existing antennas. Furthermore, the simulated and measure results indicate that the presented antennas also have simple structure, acceptable radiation patterns and these advantages make the antenna a suitable candidate for practical UWB applications.

5. References

[1] Chen, D. and C. H. Cheng 2009 *A novel compact ultra-wideband (UWB) wide slot antenna with via holes* Progress In Electromagnetics Research **94** 343-349.

[2] Lin, S., R.-N. Cai, G.-L. Huang, and J.-X. Wang 2011 *A miniature UWB semi-circle monopole printed antenna* Progress In Electromagnetics Research Letters **23** 157-163.

[3] Lin, C.C. and H.R. Chuang 2008 *A 3-12GHz UWB planar triangular monopole antenna with ridged ground-plane* Progress In Electromagnetics Research **83** 191-198.

[4] Azim, R. and M. T. Islam 2013 *Compact planar UWB antenna with band notch characteristics for WLAN and DSRC* Progress In Electromagnetics Research **133** 391-406.

[5] Akhoondzadeh-Asl, L., M. Fardis, A. Abolghasemi, and G. Dadashzadeh 2008 *Frequency and time domain characteristic of a novel notch frequency UWB antenna* Progress In Electromagnetics Research **80** 337-348.

[6] Liu, Jianjun, Karu P. Esselle, and Shun-Shi Zhong 2010 *A printed extremely wideband antenna for multi-band wireless systems* Antennas and Propagation Society International Symposium (APSURSI), IEEE.

[7] Srifi, M. N., S. K. Podilchak, M. Essaaidi, and Y. M. M. Antar 2011 *Compact disc monopole antennas for current and future ultrawideband (UWB) applications* IEEE Transactions on Antennas and Propagation **59**4470-4480.

[8] Lin, C. C., P. Jin, and R. W. Ziolkowski 2012 *Single, dual and tri-band-notched ultrawideband (UWB) antennas using capacitively loaded loop (CLL) resonators* IEEE Transactions on Antennas and Propagation **60** 102-109.

[9] Sim, C. Y. D., W. T. Chung, and C. H. Lee 2010 *Compact slot antenna for UWB applications* IEEE Antennas and Wireless Propagation Letters *9* 63-66.

[10] Bing Gong, Jin Liang Li, QiuRongZheng, Ying Zeng Yin, and Xue-Shi Ren2013 *A Compact Inductively Loaded Monopole Antenna for Future UWB Applications* Progress In Electromagnetics Research **139** 265-275.

Effects of combined apertures on the magnitudes of electric coupling coefficients of combline resonators

T M Pholele[1] and J M Chuma[2]

[1]Botswana College of Agriculture, Private Bag 0027, Gaborone, Botswana
[2]Botswana International University of Science and Technology, Private Bag 14, Phalapye, Botswana

tpholele@yahoo.com

Abstract. The effects of differently positioned and combined apertures on the magnitudes of electric coupling coefficients of two identical combline resonators are analysed. Coupling coefficient, k-vertical symmetry provides greater negative coupling for apertures less than 50 % opening on the common wall. Combined apertures do not enhance electric coupling. The Γ aperture annihilates electric coupling, whereas T shaped aperture provides reduced negative coupling coefficients compared to single apertures that make up the combination.

1. Introduction

Conventional combline filters which consist of a set of central metal posts shorted at one end and capacitively loaded at the other end of the enclosure, have been used over the years due to their low insertion loss, low production cost, compact size, relatively high unloaded quality factor and superior spurious free performance [1][2]. The modern wireless systems require miniaturised devises hence dielectric loaded combline filters have been designed and found to be significantly smaller than the equivalent air filled combline filters operating at the same resonant frequency [2]. The presence of the dielectric concentrates the electric fields in itself consequently improving the unloaded quality factor of the filter, although at the detriment of the spurious free bands which become moderate.

Coupling is very important in building filters as it is through it that we control the filter bandwidth and selectivity [3]-[7]. The selectivity of the filter mainly depends on the position of transmission zeros. These transmission zeros are brought about by cross couplings through apertures or coupling through non-resonating modes [5], [6]. The coupling can either be through electric fields, termed electric coupling, or through magnetic fields, termed magnetic coupling [7]. These couplings depend on the size and position of the apertures, the thickness of the coupling wall, the position of the resonators relative to the enclosure walls and the dimensions of the coupling screws [8]. Electric coupling generally poses challenges as it is difficult to effect.

The purpose of the study is to find out the effect of differently positioned coupling apertures (both single and combined) on the magnitudes of the coupling coefficients. A commercial software package CST Microwave studio (CST MWS) was used in the simulations after it was proven to be rigorous. It was established that single apertures provide greater electric coupling compared to combined apertures. The Γ-combined aperture annihilates electric coupling while the T-combined apertures provides a reduced negative coupling compared to the single apertures that make it up.

2. Analysis Technique

A combline resonator designed by Yao et al. [9] was implemented and run in the CST MWS and the coupling coefficients computed. To get f_e and f_m (odd and even mode resonant frequencies of coupled resonators) two identical and coupled resonators were built and the symmetry option in boundary conditions was used to assign the common wall as electric and then magnetic. The height of the common wall was milled and in each case, f_e and f_m computed. Coupling coefficient for similar resonators can be calculated using the formula 1 [10].

$$k = \frac{f_e^2 - f_m^2}{f_e^2 + f_m^2} \tag{1}$$

Figure 1 is very comparable with that obtained by Yao et al. [9]. Subject to the readability of the Yao et al. graph, the results are very close to those obtained using CST MWS as shown in Figure 2.

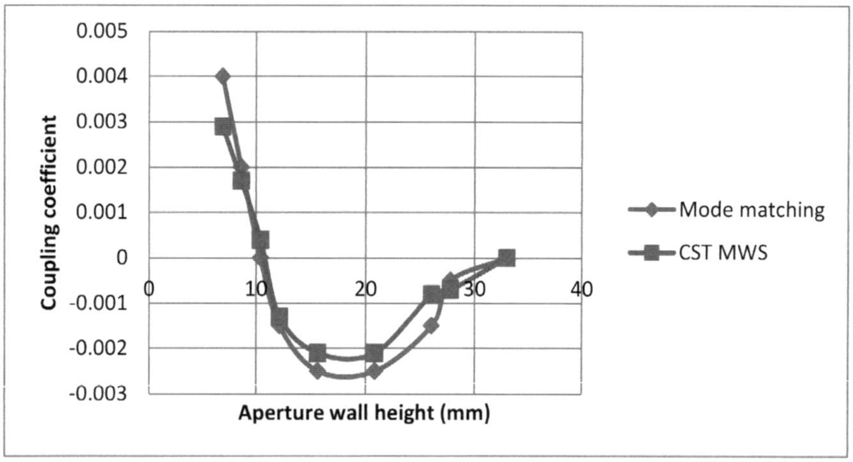

Figure 1. Coupling coefficient against change in the aperture wall height obtained using mode matching method and CST MWS.

3. Effect of size of different aperture openings on the coupling coefficient

CST MWS was used to compare the effect of differently positioned apertures on the coupling coefficients. To study the degree of coupling two identical resonators with a common wall in between were used and a method used in section 2 above was employed. The two identical coupled resonators used for the study is shown in Figure 2.

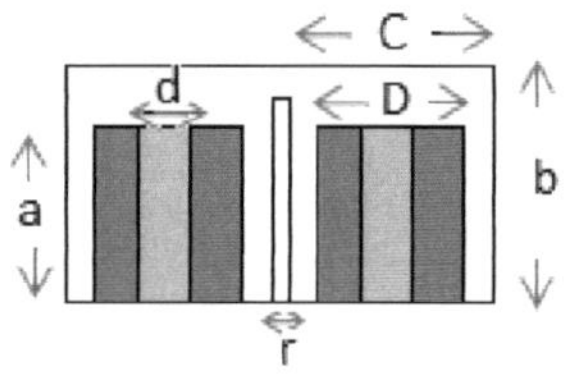

Figure 2. Cross section of coupled identical resonators; width of common wall r = 2 mm, height of cavity b = 23mm, height of both dielectric ring and combline post a = 20 mm, length and width of cavity C = 20 mm, diameter of the post d = 12.7 mm, diameter of dielectric D = 18mm. Dielectric constant = 37.

One aperture was opened horizontal from the capacitive side of the combline resonator and the coupling coefficients generated from this are labelled as k-horizontal. The other aperture studied was opened vertical from the side, and the coupling coefficients generated from this aperture were referred to as k-vertical-side. The third aperture studied was opened vertical from the centre of the common wall, and the coupling coefficients obtained from this were labelled as k-vertical-symmetry. The three cases are represented in Figure 3 as a), b) and c) respectively.

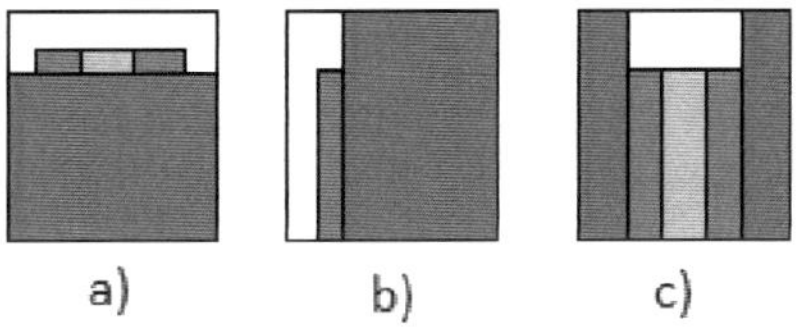

Figure 3. Cross section of horizontal, vertical-side ape and vertical-symmetric apertures respectively.

It can be observed in Figure 4 that all apertures begin with near null coupling when the apertures are very small, indicating that minimum electric and magnetic fields find their way between the two resonators. K-vertical-symmetry consistently provides superior coupling than the other two aperture positions. The k-vertical-symmetry apertures offer negative coupling for a range of apertures from about 10 % opening to just below 50 % opening on the common wall. A 0.7 ratio of the aperture of k-vertical-symmetry generates magnetic coupling which is 1.5 and 1.7 greater than k-horizontal and k-vertical side respectively. The maximum electric coupling is obtained for the three single apertures when the common walls are opened about 30 %, 40 % and 50 % for k-vertical symmetry, k-horizontal and k-vertical-side respectively. This is attributed to the fact that maximum electric field couple around this region, beyond which the magnetic field begin to dominate hence the coupling coefficients gradually become positive. The maximum electric coupling produced by k-vertical-symmetry is almost equal to that produced by k-horizontal coupling and these two are 1.19 times greater than that generated by k-vertical-side.

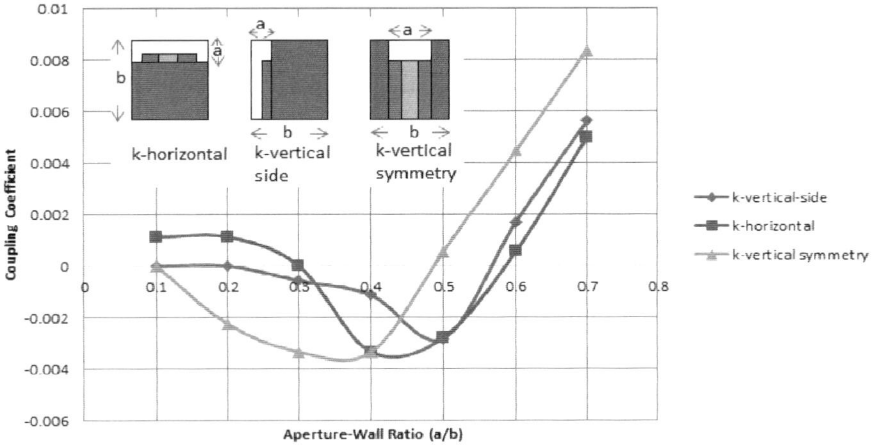

Figure 4. Coupling coefficients for aperture ratio opened horizontal, vertical-side and vertical-symmetry.

4. Combined apertures

The apertures were then milled in a combined way such that they formed the Γ shape which is a combination of the vertical-side and the horizontal aperture, and the T shaped which is a combination of the horizontal and the vertical-symmetry apertures. Simulations were run for these new apertures and f_e and f_m were obtained as was done in section 2 above and coupling coefficients were computed. The results as shown in Figure 5 show that the coupling coefficients for these combinations are not a sum of the combinations. In cases where the apertures are more than 50 % the size of the common wall, the coupling coefficients obtained are greater than the sum of the single apertures that make up the combination. This shows that these combinations do not enhance electric coupling. T shaped aperture provides a maximum negative coupling coefficient of -0.002223 which is 1.5 times less than the maximum values of the negative coupling coefficients of both the k-vertical-symmetry and k-horizontal, the single apertures that make up the combination.

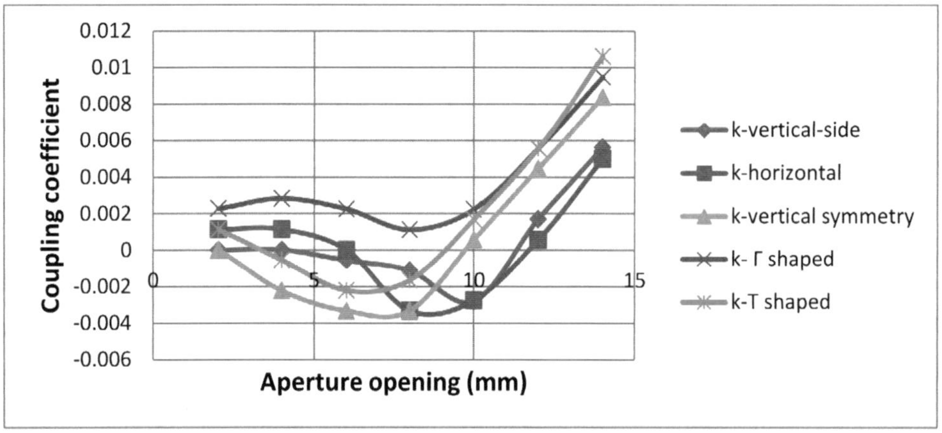

Figure 5. Coupling coefficients for single apertures; vertical-side, horizontal and vertical-symmetry, and combined apertures; Γ shaped and T shaped all against the increasing apertures.

5. Conclusion

CST MWS computed results were compared with mode matching and experimental results obtained by Yao et al. (1995) and the results found to be very comparable. The coupling coefficients of identical coupled combline resonators for k-vertical-symmetry were found to be consistently significant for apertures less than 50 % of the common wall although has the same maximum coupling coefficient as k-horizontal aperture . It was further established that the combined apertures do not enhance electric coupling.

References:

[1] Mansour R R 2004 Filter technologies for wireless base stations *IEEE Microw. Mag.* **5** 68
[2] Chuma J M and Mirshekar-Syahkal D 2000 Compact dielectric loaded combline filters with low insertion loss *Microw. Conf.*
[3] Fiedziuszko S J 1982 Dual-mode dielectric resonator loaded cavity filters *IEEE Trans. Microw Theory Tech.* **30(9)** 1311
[4] Hoft M 2010 Dielectric TE dual mode resonator filters *Proc. German Microw Conf.* 206
[5] Pelliccia L, Cacciamani F, Tomassoni C & Sorrentino R 2012 Ulta-compact filters using TM dual-mode dielectric-loaded cavities with asymmetric transmission zeros *Microw. Symp.*
[6] Lapidus A D and Rossiter C 2004 Cross-coupling in microwave bandpass filters *Microw J.*
[7] Awai A and Zhang Y 2007 Coupling coefficients of resonators- an intuitive way of its understanding *Electronics and Communications in Japan* **90(9)** 11
[8] Chuma J and Mirshekar-Syahkal D 2002 Effects of walls on coupling between coupled combline resonators *IEE Proc.-Microw. Antennas Propag.* , **149(1)** 53
[9] Yao H W, Zaki K A, Atia A E and Hershtig R 1995 Generalised slot coupled combline filters *Microw. Symp. Digest.*
[10] Zaki K A and Chen C 1987 Coupling of non-axially symmetric hybrid modes in dielectric resonators *IEEE Trans. Microw. Theory Tech.* **35(12)** 1136
[11] CST studio suite: *www.cst.com/*2012

IOP Conf. Series: Materials Science and Engineering **67** (2014) 012006 doi:10.1088/1757-899X/67/1/012006

Design and development of deflector cavity for the non-interceptive bunch length measurement system

H. K. Pandey[1, 2, #], S. Dechoudhury[1], D .P. Dutta[1], T. K. Mandi[1], T. K Bhattacharya[2] and A. Chakrabarti[1]
[1]Variable Energy Cyclotron Centre, DAE, 1/AF, Bidhan Nagar, Kolkata, India.
[2]ATDC, Indian Institute of Technology, Kharagpur, WB, India

E-mail: hkpandey@vecc.gov.in

Abstract. A non-interceptive bunch length detector system for the measurement of bunch width of accelerated beam has been designed and developed. This detector system is based on emitted secondary electrons produced by a primary ion beam hitting a thin tungsten wire placed in the beam path.The measurement of the longitudinal beam shape for wide range of beam energy, intensity as well as ion species is possible with this detector. One of the main components of the detector system is a RF deflector cavity used to deflect electrons in correlation with rf phase of the accelerator. The detailed design, development with measurement results of the deflector cavity resonator have been reported in this paper.

1. Introduction

A Radio Frequency Quadruple (RFQ) Linear accelerator of the upcoming Radio Active Ion Beam facility (RIB) at VECC has been designed and developed [1], [2]. RFQ resonating at 37.8 MHz would be the first post-accelerator which will accelerate RIBs from1.7keV/u to 100 keV/u. Further acceleration to 400 keV/u will be carried out in the subsequent Inter-digital-H type Linacs. Efficacy of acceleration in consecutive accelerators critically depends on matching the longitudinal emittance of beam to the longitudinal acceptance of the subsequent accelerator. Thus, measuring the bunch length of ion beam accelerated by RFQ is of paramount importance for accelerating beams to higher energies without particle loss. Different types of measurement techniques have been developed for diagnostic of longitudinal beam bunch length [3]. A non interceptive bunch length detector system based on secondary electron emission, produced by a primary ion beam hitting on a tungsten wire, is a more attractive and widely used system for achieving better resolution specially when the beam current and energy is low [4], [5]. In this method, the measurement is based on a coherent transformation of the longitudinal shape of the primary beam into a transverse distribution of a secondary electron beam through the phase scanning of rf applied on the deflector electrode. In order to have half wavelength resonator is a limited space many researchers [6], [7] have used the special design of the cavity by replacing the inner conductor of the coaxial resonator by helical coil. This kind of resonators gives quite reasonable Q value and other RF properties. This prompted us to choose the capacitive loaded helical resonator for our deflector system. The detailed design of the helical resonator of deflector cavity together with RF measurements developed for bunch length detector system has been reported in this paper.

2. Description of bunch length detector

The principle of the bunch length detector device is illustrated in figure 1. A negative high voltage DC biased tungsten wire is placed in the beam line. When the primary beam hits the wire, secondary electrons are emitted and they pass through the deflector gap in the rf cavity. Depending upon the rf phase many electrons get deflected and finally non deflected ones are detected by channeltron or MCP placed after a slit. The resonating frequency of the deflector cavity is same as that of the accelerator (RFQ). The deflection depends on the relative phase of the deflecting cavity with respect to RFQ. This would reflect in counts detected by the channeltron or MCP for different relative phases. Thus, this device converts the information in time domain to space domain.

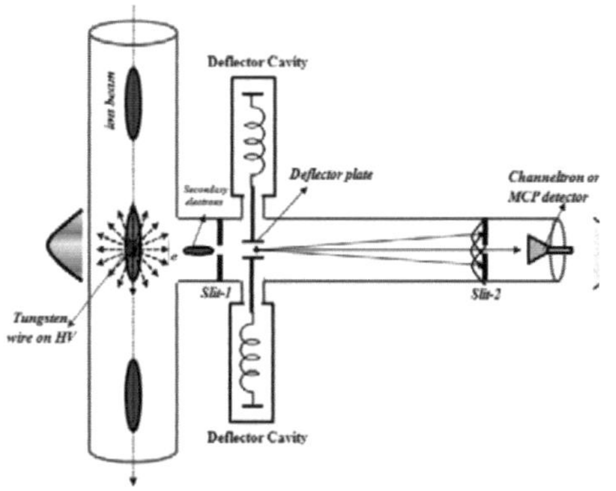

Figure 1. Schematic of the bunch length detector

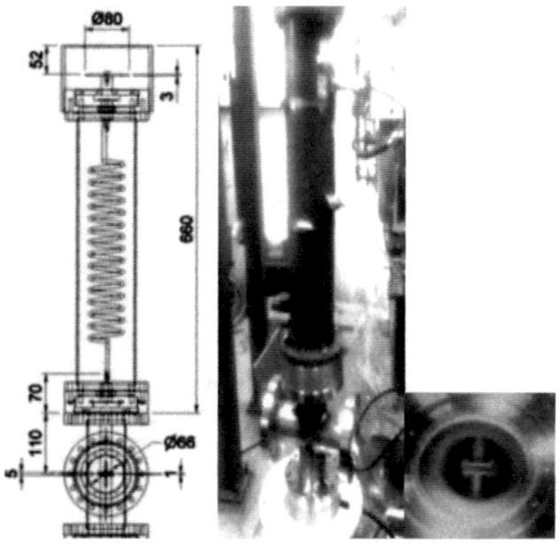

Figure 2. Deflector cavity

3. Design of deflector cavity

A DC voltage is required in the deflector cavity for efficient focusing and maximizing the count in the detector when no RF voltage is applied. So we need to develop a resonating structure where an RF signal modulates over and above a DC level. An open ended $\lambda/2$ configuration is a promising structure for our requirement. Using code SIMION [8], the geometry of the rf deflector electrode has been optimized to have desired electric field for better resolution in measuring the bunch width. The RF frequency of applied on the electrodes is 37.8 MHz, which is same as the frequency of the RFQ. To obtain the maximum deflector voltage with limited amount of driving power, a high Q resonating structure is required. A capacitive loaded half wave coaxial line resonator is a good choice. The length of an unloaded half wave coaxial resonator at an operating frequency of 37.8 MHz is 7.5 m. The resonator length can be reduced if the inner conductor of half wave coaxial line is replaced by a coil. Therefore, an alternative type of capacitance-loaded helical resonator cavity has been designed and developed. In this way we can get the reasonable Q value of the resonator and hence required deflector voltage with reduced cavity length. The resonator consists of a shielded helical coil connected to a deflector plate in one end and a tuning plate at other end as shown in figure 2. The gap between the deflector plates and the gap between the tuning plates are modeled as capacitors.The rest of the cavity is a coaxial line with helical inner conductor where the magnetic field is high. So, this part of the cavity has been modeled as an inductor. The cavity resonant frequency is given as:

$$f_o = \frac{1}{2\pi\sqrt{L_{eq} \times C_{eq}}} \tag{1}$$

IOP Conf. Series: Materials Science and Engineering **67** (2014) 012006 doi:10.1088/1757-899X/67/1/012006

where, C_{eq} is the total capacitance consisting of the deflector capacitance, tuning capacitance and self capacitance of helix of resonator. The inductance L_{eq} consists of inductance due to the optimized length of the helical line and that of the normal coaxialline which is in between helical resonator and deflector plate. The self capacitance and inductance of the helical resonator can be theoretically estimated for a given length of the helical line. The characteristic impedance of coaxial line with helical inner conductor is:

$$Z_{0H}(\Omega) = 183nd\left(\left[1-\left(\frac{d}{D}\right)^2\right]\log\left(\frac{d}{D}\right)\right)^{1/2} . \qquad (2)\#$$

Here D is the inner diameter of outer conductor (shield), d is the mean diameter of conductor in inch and n is the number of turns per inch. The use of a helical coil allows for an inductor to be made with a low self-capacitance and resistance for maximizing the Q factor. The field is more concentrated near deflector plates and the resonant frequency depends strongly on this. The fringing field effects are important input considerations during theoretical analysis and modeling of this resonator cavity. In [9], an expression for calculation of fringing field capacitance has been proposed. Using this, the capacitance of deflector plates and tuning plates have been calculated. The length of the helical resonator whose diameter of outer shield is D and mean diameter is d can be approximated for the design frequency using Eq. (2). The optimized helical length for 37.8 MHz is 330 mm with d/D ratio 0.67 and number of helical turns 17.

The required deflector voltage for desired resolution as predicted by SIMION is around 1.5 kV, which is 1.12 times of the Kilpatrick limit. The tuner plate disk diameter is 80 mm with 2mm thickness and there is a gap of about15 mm between the plates and the cavity top ceiling. The top of cavity ceiling can be moved in or out for frequency tuning. Cavity fields and Q factor have been optimized using eigenmode simulations in CST Microwave studio [10].The simulated resonant frequency, Q-value and shunt impedance are 37.83 MHz, 2000, and 10 kΩ respectively.

4. RF Measurements

We have performed the RF measurements using VNA. The measured resonance frequency is 37.801 MHz. The measured Q-value (Unloaded) using two pick-up loops under more than 98% reflection of power (-56 dB) is about 1200. The capacitive tuner can be moved over a range of 20 mm. The positions corresponding to the frequencies 37.3 MHz and 38.9 MHz are 8mm and 28mm respectively from the drift tube assembly. The measured tuning range with the gap variation of the tuning plate is shown in figure 3.The tuning range obtaining from the variation of gap from 13mm to 33mm is around 1 MHz without any appreciable change in the Q value.

The electric field pattern on the axis of buncher has been measured at low power levels using bead perturbation technique [11]. The total E_z field as a function of position along a path parallel to the z-axis has been obtained by pulling a Teflon bead of 1mm diameter through the resonator with a nylon thread. The measured E_z field for two different positions of the deflector plates is shown in figure 4.The shunt impedance measurement is necessary to determine the power requirement of cavity. The estimation of shunt impedance R_p is generally done by the capacitance variation method [12]. In this method the inter electrode capacitance of deflector is perturbed by putting a known small dielectric sheet between the plates of the deflector. R/Q is then determined as follows:

$$\frac{R}{Q} = -\frac{\Delta\omega}{\Delta C}\cdot\frac{1}{\omega_0^2} \qquad (3)$$

where ΔC is the perturbing capacitance and $\Delta\omega$ is the drift of resonant frequency due to perturbation. The estimated R_p value from the measured data is ~5.4 kΩ. Thus the maximum RF power needed for required voltage of 1.5 kV is 220 W. It is to be noted that the measured Q and R_p values are around 70% of the estimated values which are quite reasonable.

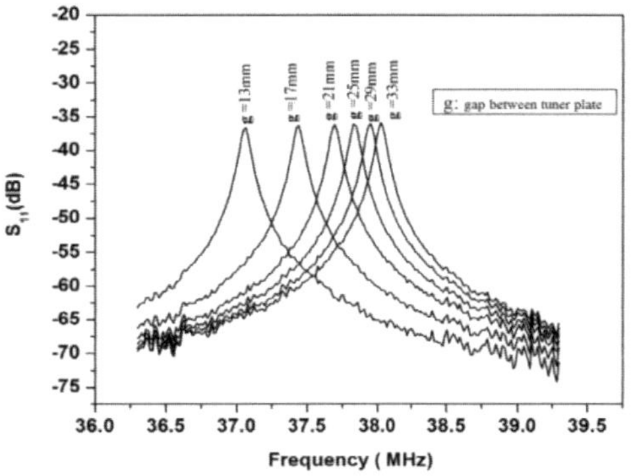

Figure 3. Change in frequency with tuner gap.

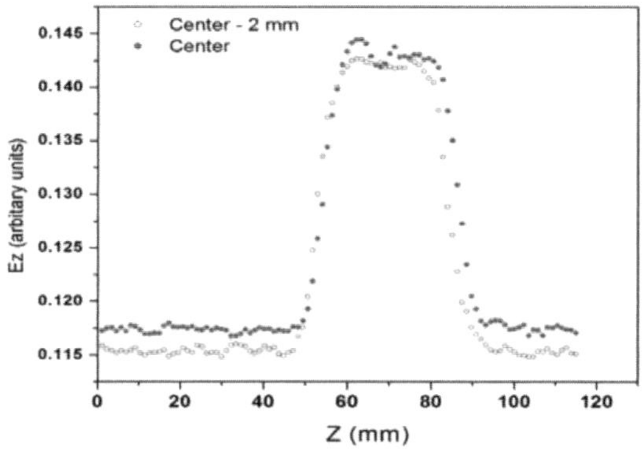

Figure 4. E_z as a function of distance along z.

5. Conclusion

A capacitive loaded helical resonator cavity for a non-interceptive bunch measurement device has been designed and fabricated. The measured RF characteristics of the resonator, such as frequency, Q value and shunt impedance have been found to be close to the analytical estimation and simulation results. In future, power coupler for RF matching and pick-up loop for feedback control will also be incorporated in the cavity. It is also planned to include a low level control mechanism in the RF system to keep the deflector voltage stable during operation.

6. Acknowledgments

The authors wish to thank the technical staff of workshop and RIB Group, VECC, Kolkata for their continuous support during the fabrication and testing of the cavity.

7. References

[1] Dechoudhury S, et. al. 2010 *Rev. Sci. Instrum.* **81, 023301.**
[2] Naik Vaishali, et. al. 2013 *Nucl. Instrum. Methods Phys .Res.* **B 317** 227.
[3] Bellato M, Dainelli A, Poggi M *Nucl. Instrum. Methods* Phys. Res. **A 382** 118.
[4] Feshenko A V *2001*, PAC-2001.
[5] Vinogradov N Y, et. al. 2002 *Proc. XXI LINAC*, Gyeongju 61.
[6] Siverns J D, Simkins L R, Weidt S, Hensinger W K 2012 Appl Phys B 107:921
[7] W. Pelzer 1992 Proc. 13th Int. Conf. on Cyclotrons and their Applications, Canada 519
[8] Dahl D SIMION-6 release, Idaho National Engineering Lab, 1996.
[9] Leus Vitaly and David Elata 2004 *TECHNICAL REPORTETR-* Israel Institute of Technology.
[10] CST Studio Suite 2008, **Version 1.2.2.** www.cst.com.
[11] Mandi T K, et. al. 2009 *Indian Particle Accelerator Conference, InPAC-2009*, RRCAT, Indore.
[12] Stokes R H et. al. 1983 *IEEE Trans. Nucl. Sci.* **NS-30** 3530.

IOP Conf. Series: Materials Science and Engineering **67** (2014) 012007 doi:10.1088/1757-899X/67/1/012007

Highly flexible short-pulse generation and high sensitivity sensing with stimulated Brillouin scattering in optical fibers

Victor Lambin Iezzi[1], Sébastien Loranger[1], and Raman Kashyap[1,2]

[1]Fabulas Laboratory, Department of Engineering Physics, Polytechnique Montreal, Montreal (Qc), Canada, H3T 1J4
[2]Fabulas Laboratory, Department of Electrical Engineering, Polytechnique Montreal, Montreal (Qc), Canada, H3T 1J4

Email: raman.kashyap@polymtl.ca

Abstract. We demonstrate two novel types of applications using the same generator as a source for Brillouin multi-wavelength generation. The first is as a highly flexible picosecond pulsed laser source which is tunable in time duration, repetition rate and emission wavelength for optical clock applications in telecommunication. The second application is a high sensitivity distributed Brillouin sensors (DTS) to lower costs and widens the market sector. We demonstrate tunability of the pulsed laser source from ~15 ps down to ~3.5 ps over the whole telecommunications C-band by simply controlling the number of Stokes waves being generated forming a phase-locked Brillouin frequency comb. The repetition rate is the Brillouin frequency shift of ~10 GHz which can be tuned by changing the gain fiber within the cavity. An increase in the standard temperature sensitivity of DTS of ~1.3 MHz/°C by 6 fold is also demonstrated. This increase is of great importance in DTS, since the detection of any variation can be made faster, which can enhance the functionality of such sensors.

1. Introduction

Brillouin scattering (BS) has for a long time been considered a nuisance in optical fiber transmission systems. Several schemes for its mitigation have been explored [1], and recently, the possibility of eliminating it altogether has been suggested by Ballato and Dragic [2]. However, there have also seen many fruitful applications of BS in optical fibers. For example, slow light generation [3], Brillouin amplifier [4], Brillouin laser cooling [5], multi-channel Brillouin laser [6] and also many types of strain and temperature sensors [7-9]. Although many types of Brillouin lasers have been proposed [10], [11], we propose a novel way to emit pulses using higher order SBS Stokes waves in, based on our first demonstration in 2012 [12]. A pulse duration in the order of a few picoseconds is achieved as well as a tunability from 1540 nm to 1565 nm is demonstrated in both the time and frequency domains. We also investigate the influence of the higher order Brillouin Stokes waves to increase the temperature sensitivity of distributed Brillouin based sensors. A cross-heterodyne detection which avoids expensive spectrum analyzer by reducing the beat frequency to near DC, is also shown.

2. Theory

Stimulated Brillouin scattering (SBS) is a well-known 3rd order nonlinear effect which is caused by the coupling and enhancement via electrostriction, from a pump wavelength and an acoustic wave (acoustic phonon) which generates a backward travelling wave in optical fiber, shifted by ~10GHz, called a Stokes wave. The frequency shift is determined by equation (1) below:

$$\nu_{B0} = 2n_p V_A / \lambda_p \qquad (1)$$

Where n_p is the effective refractive index, V_A the longitudinal acoustic phonon velocity and λ_p the pump wavelength. This Brillouin frequency shift also exhibits a linear dependence over temperature and strain which can be expressed such as:

$$\nu_B(T,\varepsilon) = \nu_{B0} + C_T(T - T_0) + C_\varepsilon(\varepsilon - \varepsilon_0) \qquad (2)$$

Where C_T is the temperature frequency coefficient, C_ε is the strain-frequency coefficient and ν_{B0} is the Brillouin frequency shift at a reference temperature T_0, and strain reference, ε_0. When cascaded, the frequency comb can be represented as for a single resonant cavity scheme (which will be explained in the next section):

$$f_{comb} = \nu_B(T,\varepsilon) + 2\nu_B(T,\varepsilon) + 3\nu_B(T,\varepsilon) + ... + n\nu_B(T,\varepsilon) \quad (3)$$

In which n represents the number of Stokes waves being generated. For a dual cavity configuration, this will be doubled ($2n$). The frequency shift of each element can be detected by beating a reference frequency comb source (reference temperature or strain) and a sensing source from a fiber experiencing temperature/strain variations. This lead to a beat frequency spectrum, in which the term ν_{B0} from equation (2) cancels out and thus the only beat signal is generated from the difference in frequencies induced by temperature or strain variations, between the two fibers. This beat note can be represented by the following equation:

$$\nu_{beat} = \nu_{ref}(T_0,\varepsilon_0) - \nu_{test}(T,\varepsilon) = \sum_n \left\{C_{n,T}\Delta T + C_{n,\varepsilon}\Delta\varepsilon\right\} \quad (4)$$

Where $C_{n,T}$ and $C_{n,\varepsilon}$ are the respective coefficient of temperature and strain dependence related to the nth Stokes order in which the 1st Stokes wave of the reference generator will beat with the 1st order Stokes of the test generator and so on, being displayed at the output by an electric spectrum analyzer.

3. Experimental setup

We proposed two cavity configurations for our applications. The first cavity has even and odd Stokes waves at the output together spaced by the natural Brillouin frequency shift of ~10GHz. The other has only the odd or even Stokes waves at the output, spaced by twice the frequency shift (~20GHz), as shown in Figure 1.

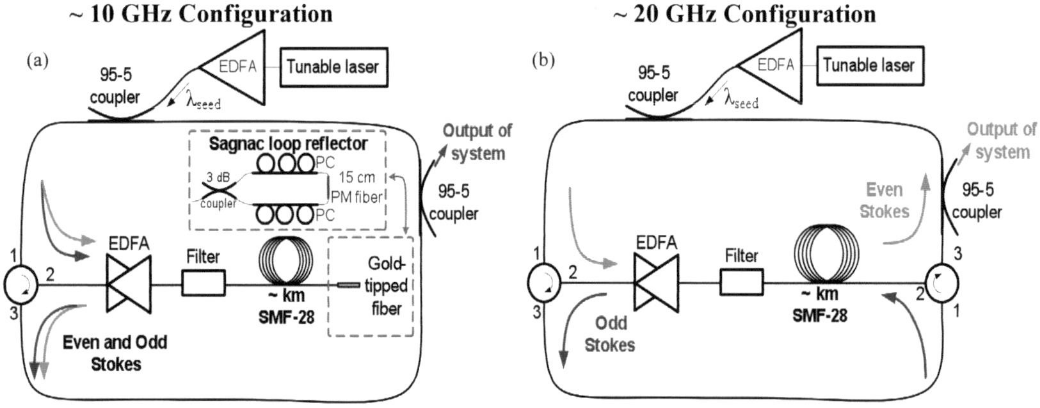

Figure 1. (a) Experimental setup for self phase-locked SBS pulse generation, for a unique cavity system using even and odd Stokes waves together and (b) the even/odd Stokes separately.

For the single cavity (figure 1(a)), a Sagnac loop interferometer can be used as a reflective mirror [13] or a gold tipped mirror to allow all the Stokes waves to travel in the same direction. As for the dual cavity (figure 1(b)), a second circulator is used to separate the odd from the even Stokes waves and depending on the position of the output coupler, the user can use either The main difference is in the repetition rate of the SBS pulsed laser, since by Fourier transform, the spacing between two frequencies determines the repetition rate of the pulse train. The total bandwidth determines the pulse duration, which in this case is unchanged.

4. Results

By controlling the power of the seed laser, and the gain of the in-cavity EDFA, it is possible to control precisely the number of Stokes waves being generated and thus the pulse duration. Figure 2(a)-(b) present this duality in which the number of Stokes waves within an arbitrary criterion of a -20 dB bandwidth are increased from 5 to 28 in the single cavity configuration, which leads to pulses changing from 15.4 ps to 3.65 ps. figure 2(c), shows the tunability on is presented over C-band covering from 1540 nm < λ< 1565 nm.

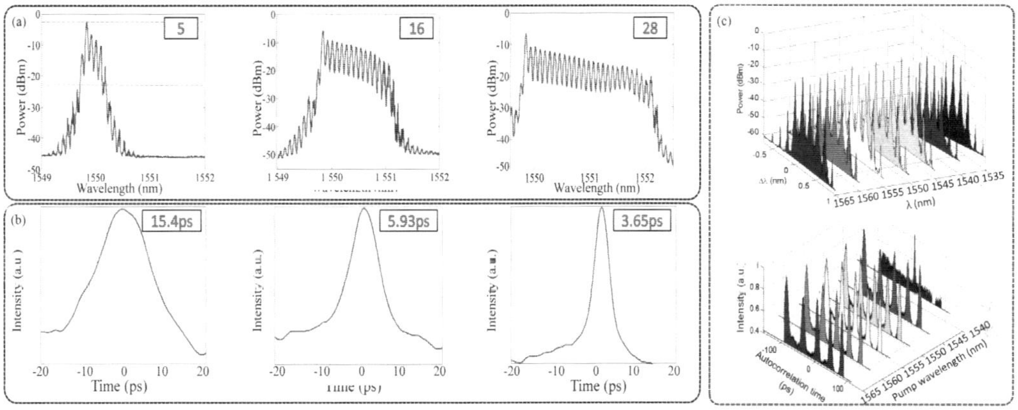

Figure 2. Demonstration of the effect of the spectral width on the pulse duration for unique cavity configuration (a) Increase, from left to right, of the number of Brillouin peaks being generated and (b) decreasing of the pulse duration in regard of the number of Stokes waves generated. (c) Wavelength tunability demonstration for the duality over a range from 1540 nm < λ< 1565 nm.

To obtain the maximum number of Stokes peaks at a given wavelength, the seed wavelength must be changed appropriately as well as the tunable filter which needs to be placed within the proper spectral region which narrows the amplification bandwidth to the width of the desired frequency comb limiting the amplified spontaneous emission generated, or the self lasing cavity modes which compete with the multi Stokes Brillouin lines. This effect can be seen in figure 2(c) in which at 1540nm in the time domain, the pulses lose their coherence since the Stokes lines are competing with the other effects present in the ring cavity.

For the temperature sensor, a reference SBS frequency comb source beats with the sensing source to achieve low frequency measurements of the frequency shift, shown in figure 3(a). The common seed pump laser for the two optical circuits removes any frequency drift which can occur after recombination of the two system outputs. Each reference Stokes is compared with its equivalent in the sensing SBS fiber, therefore giving an overlapping beat spectrum. Cross-Stokes beat frequencies are restricted observation to frequencies below 1 GHz using the low frequency electrical spectrum analyser (ESA). Figure 3(b) presents experimental results of the increase in sensitivity with the Stokes order being observed where the temperature dependence with the Stokes order is shown.

5. Discussion

The main challenge for the pulsed system is to control all of the parameters to achieve a good extinction ratio in the time domain. A predominant pedestal of roughly 40% noise in the autocorrelation trace measurements is seen. Decreasing the pulse width to sub picosecond will be the next generation of SBS pulsed laser sources. As for the sensor, generating more Stokes waves will increase the sensitivity by at least another order of magnitude, as well as for show strain sensing. Temperature sensitivity improvement for a fully distributed sensor to track a variation of temperature locally with a high resolution over a long length of fiber has to be implemented.

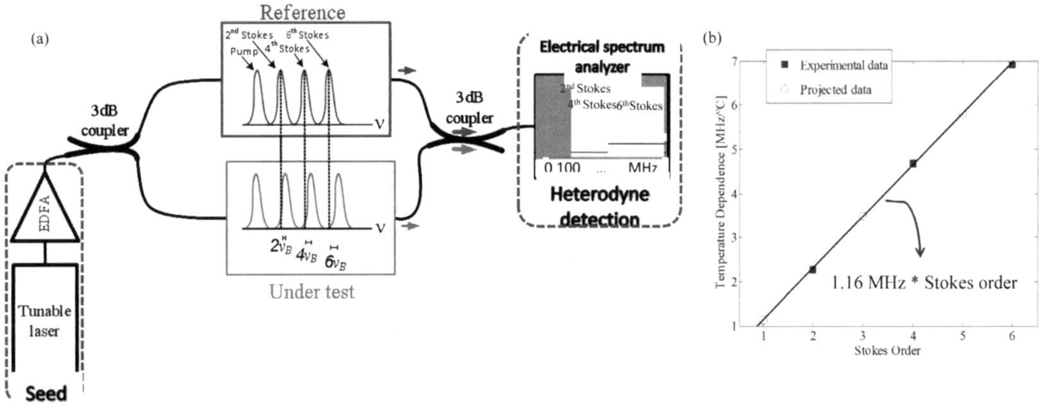

Figure 3. (a)Experimental setup to detect the increase in the temperature sensitivity of the Brillouin based sensor by self-heterodyne detection, (b) Increase of strain/temperature sensitivity with the order of the generated Stokes wave.

6. Conclusion

Two configurations of a multi-channel Brillouin generator that can either select all the Stokes waves together or only the even/odd Stokes separately have been shown. A novel type of Brillouin pulsed laser system that can be tuned to generate pulses from 15.4 ps to 3.65 ps and that can also emit at any desired wavelength at the CW seed laser has been demonstrated. In addition, a novel method is shown to increase the temperature sensitivity as a function of the generated Stokes order of a Brillouin based sensor. In the present scheme up to 6 *x* the standard sensitivity has been demonstrated.

7. Acknowledgements

RK acknowledges support from the Canada Research Chairs program and the Natural Sciences and Engineering Research Council of Canada's Discovery Grants program.VLI and SL acknowledge support from NSERC.

References

[1] Kobyakov A, Sauer M, and Hurley J. E 2005, *Optical Fiber Comm.Conf* OME5
[2] Ballato J and Dragic P 2013, *J American Ceramic Society* **96** 2675-2692
[3] González-Herráez M, Song K.-Yand Thévenaz L 2005, *Appl. Phys. Lett.* **87** 081113
[4] Miller E. J, Skeldon M. D and Boyd R. W 1989, *Appl.Opt.* **28** 92-96
[5] Bahl G, Tomes M, Marquardt F and Carmon T 2012, *Nat.Phys.* **8** 203-207
[6] Pelusi M. D, Fu A and Eggleton B. J 2010, *Opt.Exp.* ***18*** 9435-9446
[7] Bao X, Webb D. J and Jackson D. A 1994, *Opt. Lett.* **19** 141-143
[8] Hotate K and Ong S. S 2003, *IEEE Photon. Tech.Lett.* **15** 272-274
[9] Alahbabi M. N, Cho Y. T and Newson T. P 2005, *JOSA B* **22** 1321-1324
[10] Stokes L, Chodorow M and Shaw H 1982,*Opt. Lett.* **7** 509-511
[11] Picholle E, Montes C, Leycuras C, Legrand O and Botineau J. 1991, *Phys.Rev.Lett.* **66** 1454
[12] Loranger S, Lambin-Iezzi V and Kashyap R 2012, *Opt. Exp.* **20** 19455-19462
[13] Fok M. P and Shu C 2006, *Opt. Exp.* **14** 2618-2624

Superdirective all-dielectric nanoantennas: theory and experiment

P. A. Belov, A. E. Krasnok, D. S. Filonov
ITMO University, St.Petersburg 197101, Russia

C. R. Simovski
Aalto University, School of Electric and Electronic Engineering, Aalto F176000, Finland

Yu. S. Kivshar
Australian National University, Canberra ACT 0200, Australia

E-mail: belov@phoi.ifmo.ru

Abstract. We introduce a novel concept of superdirective nanoantennas based on the generation of higherorder optically-induced magnetic multipoles. Such an all-dielectric nanoantenna can be realized as an optically small spherical dielectric nanoparticle with a notch excited by a point source located in the notch. We also confirm the predicted superdirectivity effect experimentally through scaling to the microwave frequency range.

1. Introduction

Electrically small radiating systems whose directivity exceeds significantly that of a dipole are usually called superdirective [1]. Superdirectivity is an important property of radio-frequency antennas employed for space communications and radioastronomy, and it can be

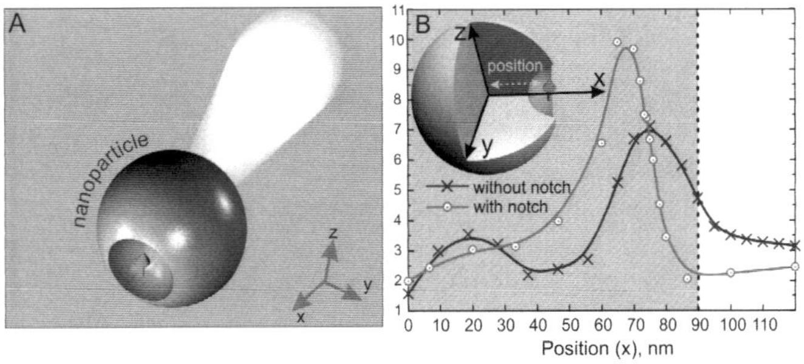

Figure 1. (**A**) Geometry of the notched all-dielectric nanoantenna. (**B**) Maximum of directivity depending on the position of the dipole ($\lambda = 455$ nm) in the case of a sphere with and without notch, respectively.

achieved in antenna arrays in a narrow frequency range and for a sophisticated system of phase shifters [1]. Achieving high radiation directivity is also important for actively studied optical nanoantennas [2, 3]. Usually, small plasmonic nanoantennas possess weak directivity close to that of a point dipole. Despite its importance, antenna's superdirectivity in the optical frequency range was not discussed or demonstrated so far. Recently, it was suggested to use different dielectric and semiconductor materials for the development of antennas at the nanoscale [4, 5, 6]. Such all-dielectric nanoantennas consist of high-permittivity nanoparticles having both resonant electric and magnetic optical responses [4, 5, 6]. This approach allows to study an optical analogue of the so-called Huygens source, an elementary emitting system with properly balanced electric and magnetic dipoles oscillating with the same phase [1, 4, 5, 6]. As a result, nearly twice higher directivity than that of a single electric dipole has been reported. However, this directivity is still insufficient for nanophotonics applications.

2. Results and discussions

Here we reveal a way for achieving superdirectivity of antennas with a subwavelength (maximum size 0.4-0.5 λ) radiating system without using complex antenna arrays, and it is valid for a wide range of frequencies. First, we demonstrate possibility to create a superdirective optically small nanoantenna that does not require metamaterial. We consider one semiconductor nanoparticle with the permittivity $\mathrm{Re}\varepsilon = 15 - 16$ radiated by light at wavelength λ (for λ =440-460 nm this corresponds to a nanoparticle made of crystalline silicon [9]) and the radius $R_{\mathrm{s}} = 90$ nm being almost five times smaller than λ. For a perfect sphere, lower-order multipoles for both electric and magnetic fields are excited while the contribution of higher-order modes is negligible [4, 5, 6, 8]. However, making a small notch in the spherical particle breaks the symmetry allowing the excitation of higher-order multipole moments of the sphere. This is achieved by placing a nanoemitter (e.g. a quantum dot) within a small notch created on the sphere surface, as shown in Fig. 1A. The notch in our example has the shape of a hemisphere with a radius $R_{\mathrm{n}} \ll R_{\mathrm{s}}$. The emitter can be modeled as a point-like dipole and it is shown in the figure by a red arrow. It turns out that such a small modification of the sphere would allow the efficient excitation of higher-order spherical multipole modes.

To study the problem numerically, we employ the software package CST Microwave Studio. To characterize the antenna properties we calculated the directivity D defined as:

$$D = \frac{4\pi \mathrm{Max}[p(\theta, \varphi)]}{P_{\mathrm{rad}}}, \tag{1}$$

where $P_{\mathrm{rad}} = \int_{\Omega} p(\theta, \varphi) d\Omega$ is a total power radiated in the far field, $\mathrm{Max}[p(\theta, \varphi)]$ is the power flux density radiated in the direction of maximal flux density, θ and φ are the angles of conventional spherical system, $d\Omega$ is the elementary solid angle. Figure 1B shows the dependence of the maximum directivity D_{mav} on the position of the emitting dipole in the case of a sphere $R_{\mathrm{s}} = 90$ nm without a notch, at the wavelength $\lambda = 455$ nm (blue curve with crosses). This dependence has the maximum ($D_{\mathrm{max}} = 7.1$) when the emitter is placed inside the particle at the distance 20 nm from its surface. The analysis shows that in this case the electric field distribution inside a particle corresponds to the noticeable excitation of higher-order multipole modes. This becomes possible due to strong inhomogeneity of the external field produced by the nanoemitter. Furthermore, the excitation of higher-order multipoles can be significantly improved by making a small notch in the silicon spherical nanoparticle and placing the emitter inside that notch, as shown in Fig. 1. This modification of the nanoparticle transforms it into a resonator for high-order multipole moments. In our problem, the notch has the form of a hemisphere with the center it the dielectric nanoparticle's surface. The optimal radius of the notch is $R_{\mathrm{n}} = 40$ nm, that we find by means of numerical optimization. Red curve with circles in the Fig. 1b shows maximum of directivity corresponding to this geometry. Maximal directivity at wavelength 455 nm is $D_{\max} = 10$.

Expansion of the field into multipoles offers an illustrative description of the internal field composition for the spherical particle. This is given by a series of spherical harmonics with the coefficients $a_E(l, m)$ and $a_M(l, m)$, which characterize the electrical and magnetic multipole moments [10, 7].

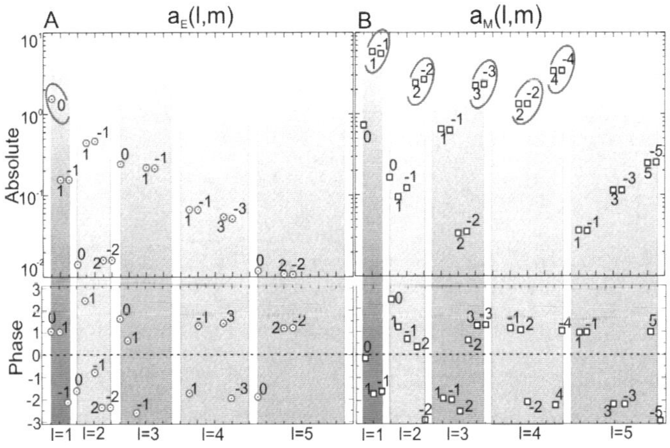

Figure 2. Absolute values and phases of (**A**) electric and (**B**) magnetic multipole moments that provide the main contribution to the radiation of all-dielectric superdirective optical nanoantenna at the wavelength 455 nm. Multipole coefficients providing the largest contribution to the antenna direction are highlighted by red circles. Particle and notch radii are equal to $R_s = 90$ nm and $R_n = 40$ nm, respectively.

In general, the multipole coefficients determine not only the mode structure of the internal field but also the angular distribution of the radiation. Internal field was calculated numerically with the expansion into the multipole series. The multipole moments for this case are shown in figure (Fig. 2), where we observe strong excitation of $a_E(1, 0)$, $a_M(1, 1)$, $a_M(1, -1)$, $a_M(2, 2)$, $a_M(2, -2)$, $a_M(3, 3)$, $a_M(3, -3)$, $a_M(4, 2)$, $a_M(4, -2)$, $a_M(4, 4)$ and $a_M(4, -4)$. These multipole moments determine the angular pattern of the antenna. All other ones give a negligible contribution. Absolute values of all magnetic moments are larger than those of the electric moments in the corresponding multipole orders, and the effective spectrum of magnetic multipoles is also broader than that of the electric moments. Thus, the operation of the antenna is mainly determined by the magnetic multipole response. Absolute values of multipole coefficients $a_M(l, \pm|m|)$ of the same order l are practically equivalent, however, the phase of these coefficients are different. Therefore, the modes with $+|m|$ and $-|m|$ form a strong anisotropy of the forward–backward directions that results in the unidirectional radiation.

We have confirmed both predicted effects studying the similar problem for the microwave range. To do this, we have scaled up the nanoantenna as above to low frequencies. Instead of Si we employ MgO-TiO_2 ceramic [5] characterized at microwaves by a dispersion-less dielectric constant 16 and dielectric loss factor of $1.12{\cdot}10^{-4}$. The results of the experimental investigations and numerical simulations of the pattern in both E- and H-planes are summarized in (Figs. 3A,B). Radiation patterns in both planes are narrow beams with a lobe angle about 35°. Experimentally obtained coefficients of the directivity in both E- and H-planes are equal to 5.9 and 8.4, respectively (theoretical predictions for them were equal, respectively, 6.8 and

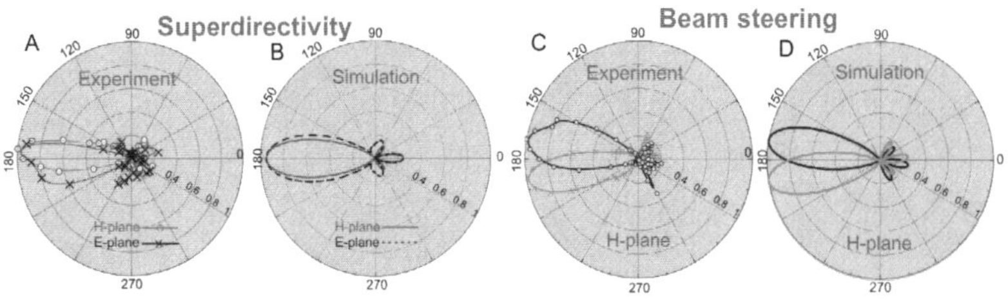

Figure 3. (color online) Experimental (**A**) and numerical (**B**) radiation patterns of the antenna in both E- and H-planes at the frequency 16.8 GHz. The crosses and circles correspond to the experimental data. Experimental (**C**) and numerical (**D**) demonstration of beam steering effect, displacement of dipole is equal 0.5 mm.

8.1). Our experimental data are in a good agreement with the numerical results except a small difference for the E plane, that can be explained by the imperfect symmetry of the emitter. Note, that the observed directivity is close to that of an all-dielectric Yagi-Uda antenna with overall size 2λ [5]. The total size of our experimental antenna is closed to $\lambda/2.5$. Thus, our experiment clearly demonstrates the superdirective effect.

Experimental and numerical demonstration of the beam steering effect are presented in (Figs. 3C,D). For the chosen geometry of antenna, displacement of source by 0.5 mm leads to a rotation of the beam about 10°. Note that the ratio of $\lambda = 18.7$ mm to value of the source displacement 0.5 mm is equal to 37. This proves that the beam steering effect observed at subwavelength displacement of source.

Acknowledgments
This work was supported by the Ministry of Education and Science of the Russian Federation (GOSZADANIE 2014/190, Zadanie No. 3.561.2014/K), Dynasty Foundation (Russia), and the Australian Research Council.

References

[1] C. Balanis, *Antenna Theory: Analysis and Design.* New York ; Wiley, 1982.
[2] A. Alu and N. Engheta, *Phys. Rev. Lett.*, vol. 104, pp. 213902, 2010.
[3] L. Novotny, N. van Hulst, *Nat. Photon.*, vol. 5, pp. 83, 2011.
[4] A. E. Krasnok, A. E. Miroshnichenko, P. A. Belov, Yu. S. Kivshar, *JETP Lett.*, vol. 94, 635, 2011.
[5] D. S. Filonov, A. E. Krasnok, A. P. Slobozhanyuk, P. V. Kapitanova, E. A. Nenasheva, Yu. S. Kivshar, P. A. Belov, *Appl. Phys. Lett.*, vol. 100, 201113, 2012.
[6] A. E. Krasnok, A. E. Miroshnichenko, P. A. Belov, Yu. S. Kivshar, *Optics Express*, vol. 20, 20599, 2012.
[7] Alexandr Krasnok, Constantin Simovski, Pavel Belov, and Yuri S. Kivshar, *Nanoscale*, vol. 6, 7354-7361, 2014.
[8] Alexander E. Krasnok, Dmitry S. Filonov, Constantin R. Simovski, Yuri S. Kivshar, and Pavel A. Belov, *Appl. Phys. Lett.*, vol. 104, 133502, 2014.
[9] E. Palik, *Handbook of Optical Constant of Solids.* San Diego, Academic, 1985.
[10] J.D. Jackson, *Classical Electrodynamics.* New York : Wiley, 1998.

IOP Conf. Series: Materials Science and Engineering **67** (2014) 012009 doi:10.1088/1757-899X/67/1/012009

Impact of the coupling effect and the DC-combining configuration on a compact rectenna array

J Rivière, A Douyere and J D Lan Sun Luk

Energy, Electronics and Processes Laboratory, La Reunion University, 97744 Saint Denis, Cedex 9, France.

Email: jeromc.rivicrc@univ-rcunion.fr

Abstract. This paper proposes an experimental study of the coupling effect of a rectenna array. The rectifying antenna consists of a compact and efficient rectifying circuit in a series topology, coupled with a small metamaterial-inspired antenna. The measurements are investigated in the X plane on the rectenna array's behavior, with series and parallel DC-combining configuration of two and three spaced rectennas from 3 cm to 10 cm. This study shows that the maximum efficiency is reached for the series configuration, with a resistive load of 10 kΩ. The optimal distance is not significant for series or parallel configuration. Then, a comparison between a rectenna array with non-optimal mutual coupling and a more traditional patch rectenna is performed. Finally, a practical application is tested to demonstrate the effectiveness of such small rectenna array.

1. Introduction

With the development of very low-power wireless system, numerous researches have focused their attention on the feasibility of powering devices through the harvesting of ambient electromagnetic energy [1]. The main challenges in far field nondirective powering are a result of the low and variable power densities available at the receiving antenna.

Therefore, the study deals with the design, optimization and experimental characterization of a compact metamaterial-inspired antenna coupled with a rectifier. Then, to increase DC power over the load, the proposed rectenna has been interconnected to form arrays [2]. This is called DC-combining configuration where the outputs of each rectenna are combined in series, parallel or a hybrid manner. It has the advantages to ease the construction of the rectenna array oppositely to the RF-combining configuration. In RF-combining structure, the RF outputs of each antenna are combined before the rectifier circuit. The RF-combining configuration implies the use of sub-circuit to deal with power leakage and phasing between the antennas. This issue is eliminated when DC-combining is used, but the drawbacks of such configuration are lower RF-to-DC conversion and the use of several rectifying elements.

In this paper, we investigate the rectenna array's behavior using electrically small antenna. First, we inquire the effect of mutual coupling on two different DC-combining configurations, series and parallel for a finite number of elements. Then, a comparative study is conducted between a rectenna array and a single rectenna using a conventional patch antenna [3], with the same physical surface area. Last, an application of the rectenna array is tested, where a battery-less temperature sensor is powered over a distance of one meter.

2. Antenna design

The antenna used in this paper, as shown in figure 1, is a modified version of an antenna first introduced by Ning Zhu and Ziolkowski [4]. This Near Field Resonant Parasitic (NFRP) antenna is designed for ISM band and may be characterized as an electrically small antenna by a weighting factor $ka = 0.976$, where k is the wave number and a is the radius of the minimum size sphere that encloses the antenna. Moreover, this antenna naturally rejects the first harmonics, which allows the suppression of the pass-band filter at low power level. Indeed, due to the low power input, the effects of higher order harmonics are also negligible. This antenna is also referred as filtenna (filtering-antenna) and allows the size reduction of the overall rectenna to 11*49 mm2 (see figure 3).

The measured return loss parameter from 1 GHz to 5 GHz of the NFRP antenna is presented in figure 2. The antenna presents a dual-frequency resonance (1.9 GHz and 2.45 GHz). It should be noted that the simulations were performed at 2.54 GHz due to experience feedbacks on offsets between simulations and practice. Indeed, when we simulate with Momentum in Agilent ADS software, the simulation uses an infinite substrate, in contrast to the realization, where the antenna's dimensions are finite.

An experimental characterization of the antenna gain at a distance of 1 meter is conducted at 2.45 Ghz, using a bi-quad antenna with a gain $G_t = 10$ dBi as emitter source. The transmitted power is $P_t = +20$ dBm. The measured collected power at the receiving antenna is $P_r = -17.5$ dBm. The Friis equation (1) is used to calculate the gain of the antenna with $\gamma = 0.28$.

$$G_r = \frac{P_r \cdot (4 \cdot \pi)^2 \cdot D}{P_t \cdot G_t \cdot \lambda^2 \cdot \gamma} \tag{1}$$

The calculated gain $G_r = -1.74$ dBi is negative due to the compactness of the antenna. This bad gain is a common compromise when using electrically small antenna.

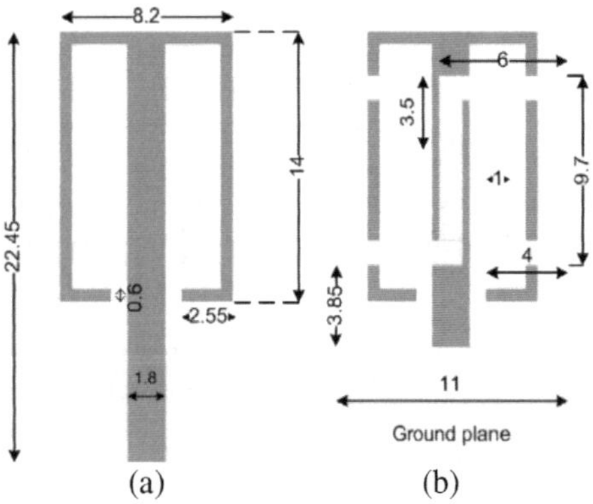

Figure 1. Small antenna design (mm)
(a) Front view (b) Back view

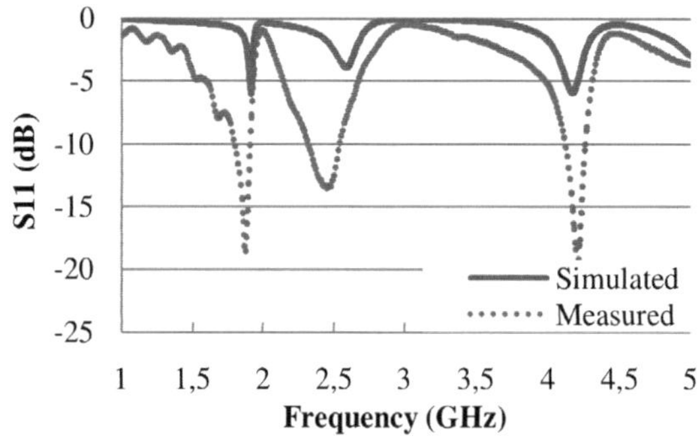

Figure 2. Simulated and measured return loss parameter

3. Rectifier design

The rectifier presented in a previous work [3] is simple, compact and efficient for low power input. A matching circuit is essential in providing the maximum power transfer from the antenna to the rectifier circuit. The optimization of the impedance matching consists of adjusting the length $L1$ of a shorted stub and the length $L2$ of a transmission line. Given the global rectifying circuit at $P_{collected} = -20$ dBm with $R_{load} = 3.5$ kΩ, and using tuning method under Momentum in ADS software, the optimized lengths values are: $L1 = 2.9$ mm and $L2 = 10.9$ mm. The rectifier circuit shows an RF/DC conversion efficiency of approximately 20% for a collected power of 10 μW. The antenna and the rectifier are realized on the same unit substrate (AD320) with $\varepsilon_r = 3.2$, $thickness = 0.762$ mm (see Figure 3).

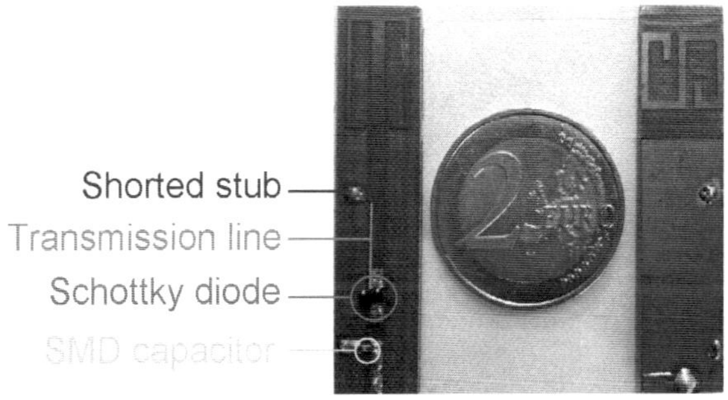

Figure 3. Picture of the global rectenna

4. Measurement and application

4.1. Experimental setup and measurements protocol

After the validation of the rectifying circuit and the NFRP antenna separately, several rectennas are realized individually and DC-combined in series or parallel. The rectennas are measured in free space and the measurement bench is shown in figure 4. A linearly polarized patch antenna with a gain approximately of 6 dBi provides the RF signal to the rectenna array.

The purpose of this study is to experimentally characterize the optimal distance between the rectennas to form an efficient array. Two rectennas and three rectennas are placed in the X plane, separated from 3 cm to 10 cm. For every distance, the DC-combining series and parallel are investigated. The voltage and current of every case are measured and averaged (100 measurements spaced of one second) at a distance of 1 meter, with an injected power $P_{inj} = +20$ dBm. Measurements are performed with a precision of 0.0035%.

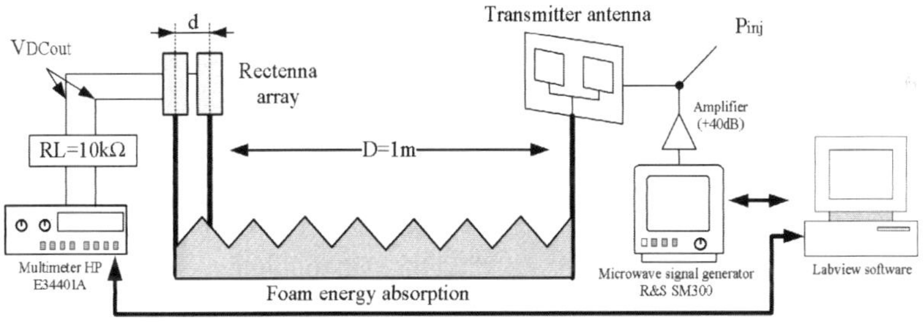

Figure 4. Measurement bench setup

4.2. Results, comparison and application

The figure 5 shows the collected DC voltage at several distances between the rectennas for each configuration. It can be noted that, despite the coupling effect not being optimal for smaller distances between the rectennas, the converted DC power is improved compared with one single rectenna. From this observation, a more compact rectenna array is build. This rectenna array is composed of 5 rectennas spaced of 9 mm, and has the same area size as a rectenna design in a previous work [3] for a comparative study (see figure 6). This previous rectenna was build for the same operating point and measured in a similar approach. Experimentally, with a resistive load $R_{load} = 10$ kΩ and the same measurement protocol, we obtain for the patch rectenna $VDC_{patch} = 226$ mV and $VDC_{small} = 1.03$ V for the DC-combining array in series configuration. The series configuration was chosen, due to its advantages to increase the voltage and being more suitable for higher resistive load.

Then, an experimental application is conducted to power a temperature sensor with a high input impedance $Z > 10$ kΩ. Due to its high input impedance, we provide enough power with the DC-combined array in series to power in continuous the sensor at a distance of one meter, as seen in figure 7. The minimum voltage needed to turn on the sensor is $V_{sensor} = 1.37\text{V}$.

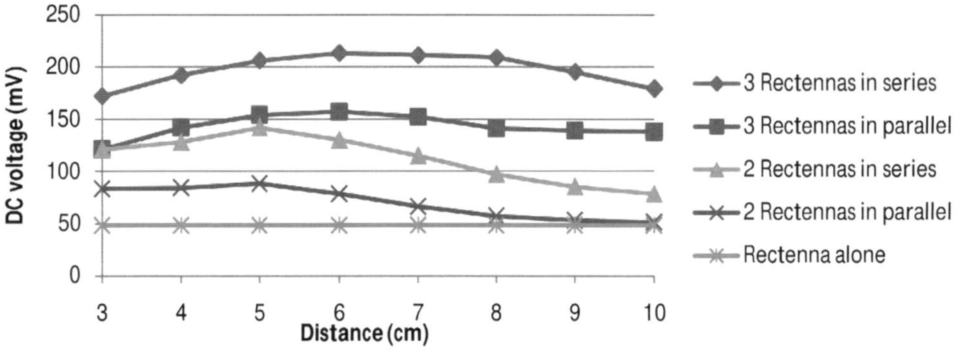

Figure 5. DC voltage collected at several distances for each configuration

Figure 6. Size comparison between regular patch rectenna and small rectenna array

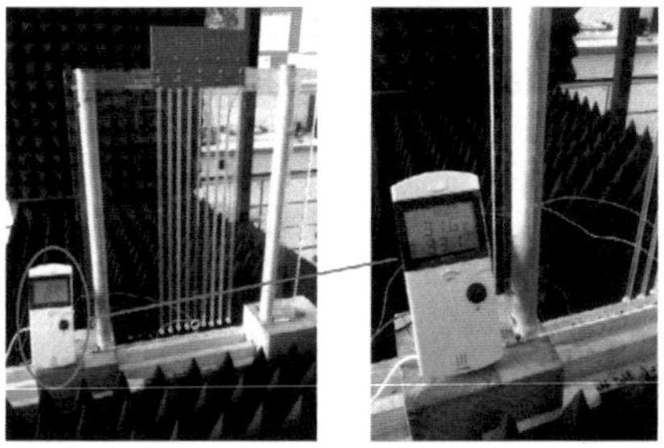

Figure 7. Wireless powering of a temperature sensor using DC-combined configuration with electrically small antenna

5. Conclusion

In this paper, a rectenna array with electrically small antenna was build using DC-combined configuration and compared with a conventional patch rectenna. We demonstrated experimentally that despite their bad gain and the effect of non-optimal mutual coupling, those arrays were useable to harvest RF energy. An application of such rectenna array has been studied. Depending on the distance and the power of the emitted source, it is possible to power constantly or periodically (using some storing elements) a low-power consumption device. Indeed, subsequently, we wish to power the same sensor at greater distance, by adding a pump charge circuit with the rectenna array.

References

[1] Mitcheson P D, Yeatman E M, Rao G K, Holmes A S and Green T C 2008 Energy Harvesting From Human and Machine Motion for Wireless Electronic Devices, *Proc. of the IEEE* vol 96 pp 1457-86

[2] Olgun U, Chi-Chih C and Volakis J L 2011 Investigation of Rectenna Array Configurations for Enhanced RF Power Harvesting *IEEE Antennas and Wireless Propagation Letters* vol **10** pp 262-5

[3] Douyere A, Alicalapa F, Riviere S and Lan Sun Luk J D 2012 Experimental study of rectenna coupling at low power level *IOP Conferences series: Materials Science and Engineering (RADIO2012)* vol 44

[4] Ning Z, Ziolkowski R W 2012 Metamaterial inspired GPS L1 rectennas *IEEE International Workshop on Antenna Technology (iWAT)* pp 193-6

IOP Conf. Series: Materials Science and Engineering **67** (2014) 012010 doi:10.1088/1757-899X/67/1/012010

Integrated RF-DC converter and PCB antenna for UHF wireless powering applications

L Vincetti[1], M Maini[1], S Scorcioni[2], L Larcher[2], A Bertacchini[2], and A Tacchini[2-3]
[1] Department of Engineering "Enzo Ferrari", University of Modena and Reggio Emilia, via Vignolese 905, 41125 Modena, Italy
[2] Department of Sciences and Methods for Engineering, University of Modena and Reggio Emilia, via Amendola 2, 42122 Reggio Emilia, Italy
[3] Fondazione REI, via Sicilia 31,42122 Reggio Emilia, Italy

E-mail: luca.vincetti@unimore.it

Abstract. In this work, a broadband differential RF-DC CMOS converter realized in CMOS 130 nm technology with a customized PCB antenna with inductive coupling feeding for RF energy scavenging is presented. Experimental results show that output DC voltage higher than 1V from 800MHz to 970MHz can be obtained with a load of 1kΩ.

1. Introduction

Scavenging the electromagnetic energy radiated in the ISM band [1], [2] is a fascinating option for the development of remote power supply solutions for ultra-low power devices like, bionic implants [3] WSN nodes [4], [5] and RFID active tags.

Rectification circuits of the RF-EH systems must be optimized to reduce the minimum power-threshold needed for the system to operate. Usually to get that target, chips containing rectifier circuits exhibit a complex input impedance with a small resistance, and a high capacitive reactance. Since matching networks cannot be used in order to minimize power loss, an impedance matching between chip and antenna is required. Broadband antennas play a key role in the RF energy harvesting if the frequency of the electromagnetic waves impinging on the system is not known *a priori*. Moreover broadband allows to compensate for the frequency shift due to the presence of unknown dielectric materials nearby the antenna.

In this work we present the results of an energy harvester module composed by a balanced RF-DC power converter and a matched broadband printed antenna [6]. No matching networks are used since the antenna is designed to match the highly capacitive chip input impedance [7]. The overall performances of a prototype have been measured in an anechoic chamber. Results show that output DC voltage higher than 1V from 800MHz to 970MHz can be obtained with a load of 1kΩ.

2. System

The RF energy harvester we developed is shown in Fig. 1. It is composed by the following block: custom differential antenna; reconfigurable RF-DC voltage rectifier (VR); the voltage regulator; the energy storage capacitor. By combining the reconfigurable VR and the custom antenna avoids the use of a dedicated matching network benefiting the efficiency of the whole system. The voltage rectifier is based on a differential reconfigurable architecture which improves the efficiency and the sensitivity of

the circuit on a very large input power range compared to typical solutions [7]. The voltage rectifier is comprised of two rectifier blocks that can be connected either in series or in parallel through the M_X, M_Y and M_Z switches, which are controlled by the control logic signal V_{CTRL}. The two modular VR blocks are derived from the classical 2-stages Dickson voltage rectifier/multiplier [8].
The antenna is composed by a printed biconical radiating dipole inductively coupled with the chip by means of a rectangular loop. The real and imaginary parts of the input impedance can be adjusted independently by acting on the loop size and its distance from radiating dipole [9]. This makes very easy to achieve the target value of Z_{ANT}. Fig. 2 shows the layout of the antenna.

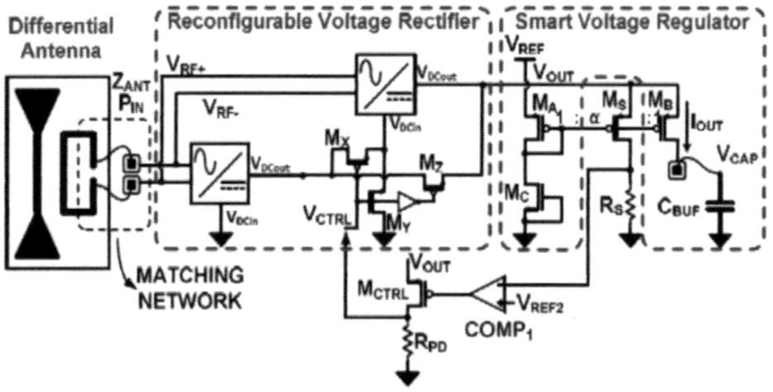

Figure 1. RF energy harvester module composed by custom differential antenna, two differential voltage rectifiers, the voltage regulator and the energy capacitor.

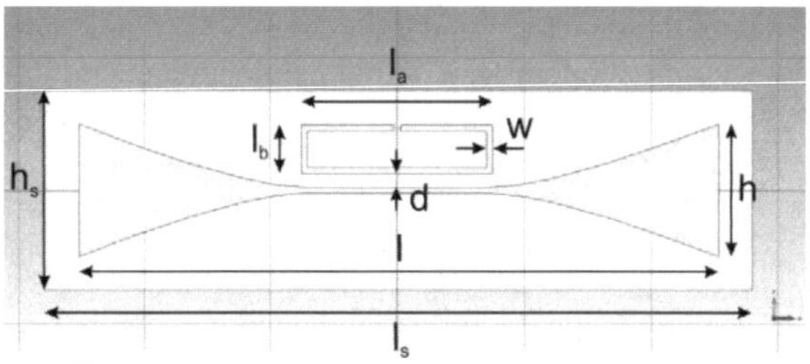

Figure 2. Layout of the differential antenna, realized on a FR4 substrate. Geometrical parameters of the antenna are reported in Table I.

The maximum power transfer from antenna to chip is obtained when the complex matching condition: is satisfied:

$$Z_{ant} = Z^*_{chip}$$

being Z_{ant} and Z_{chip} the input impedances of the antenna and of the chip respectively.
The above condition makes very challenging the design of broadband antennas because usually the input chip impedance is highly capacitive and frequency dependent. In fact, as demonstrated in [10], the rectification circuit exhibits an input impedance that can be represented as the parallel of a capacitance C_{IN} and a resistance R_{IN}. C_{IN} accounts for the rectifier transistors parasitic capacitance, while the active power flowing into the rectification circuit depends on R_{IN}.
In the chip input impedance model shown in Fig. 3 also the bond wire parasitic resistance R_{BOND} and inductance L_{BOND} are included. The antenna has been designed by considering R_{IN}=10kΩ, C_{IN}=500fF, R_{BOND}=2Ω and L_{BOND}=1nH and a substrate of FR4 with relative dielectric permittivity ε_r=4.3 and thickness of 0.8mm. The electromagnetic software CST Microwave Studio has been used for the design. The geometrical parameters of the designed antenna are reported in Table 1. The numerical values and the measured ones of the antenna input impedance $Z_{ant} = R_{ant} + jX_{ant}$ are reported in Fig. 4 and they are compared with Z^*_{chip}. The changing of the slope of X_{ant} around the resonance frequency of the radiating body f=900MHz allows to follow the conjugate chip reactance over a broad frequency

range. The antenna resistance crosses the chip resistance at f=840MHz and is higher than R_{chip} at f=900MHz. This does not allow to obtain a perfect matching condition at a given frequency but broadens the antenna bandwidth [3].

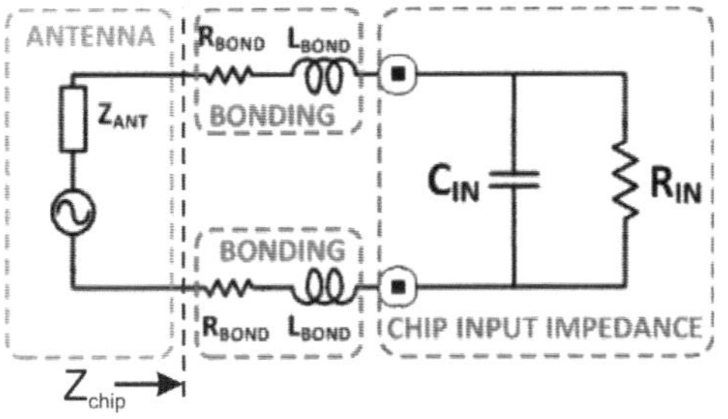

Figure 3. The equivalent circuit of the chip and antenna.

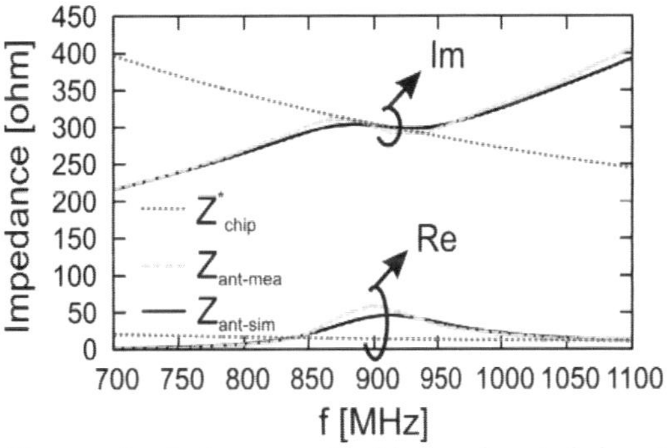

Figure 4. Antenna and chip conjugated input impedances.

Table I. Geometrical antenna parameters.

Parameter	Value
l_a	30 mm
l_b	20 mm
d	2 mm
w	1 mm
l	101 mm
h	20 mm
l_s	110 mm
h_s	30 mm

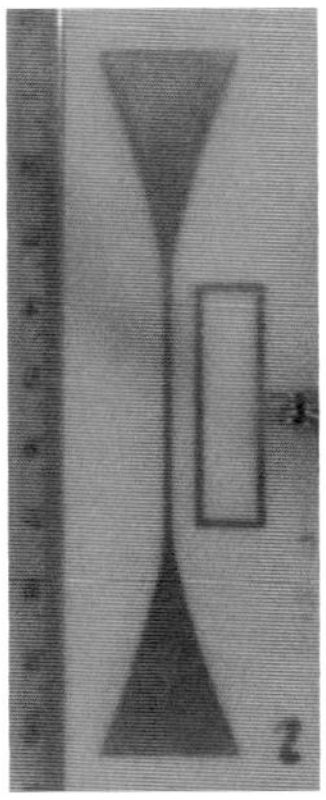

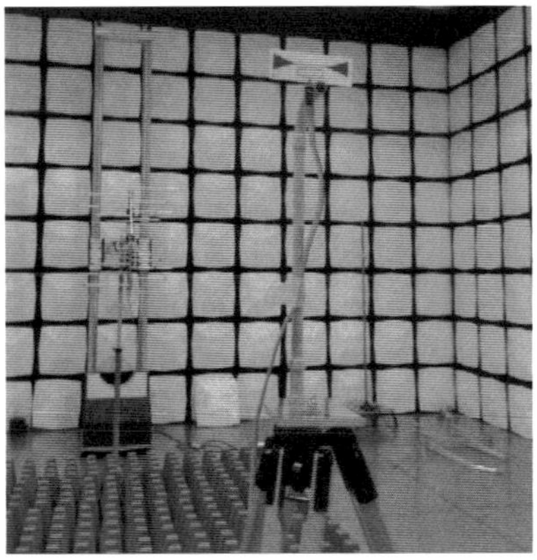

Figure 5. Left: Photograph of the PCB differential antenna realized on a FR4 substrate with bonded chip. Right: Measurement set-up.

3. Experimental results

Up to present the performances of the antenna and IC converter have been measured separately. In particular an efficiency peak of 60% has been measured for the chip [10]. Here, to test the overall

performances of the energy harvester module, a prototype of the antenna with bonded chip has been realized in it is shown in Fig. 5 (left). The performances have been measured in an anechoic chamber. The set-up is shown in Fig. 5 (right). The module has been connected to a resistive load R_L and fed with a log periodic antenna in the frequency range 800-1000MHz. The impinging electric field has been measured at the receiving antenna position in the absence of it through a probe. Results are shown in Fig. 6. They show the DC voltage on R_L versus the frequency for different values of the load and the impinging electric field. With R_L=100kΩ a DC voltage higher than 1V is obtained both at 868MHz (RFID UHF UE standard) and 920MHz (RFID UHF USA standard). With R_L=10kΩ a voltage higher than 1V is obtained over a frequency range wider than 150MHz, from 800MHz to 970MHz showing the very broadband behaviour of the developed module.
Results are shown in Fig. 6. They show the DC voltage on R_L versus the frequency for different values of the load and the impinging electric field. With R_L=100kΩ a DC voltage higher than 1V is obtained both at 868MHz (RFID UHF UE standard) and 920MHz (RFID UHF USA standard). With R_L=10kΩ a voltage higher than 1V is obtained over a frequency range wider than 150MHz, from 800MHz to 970MHz showing the very broadband behaviour of the developed module.

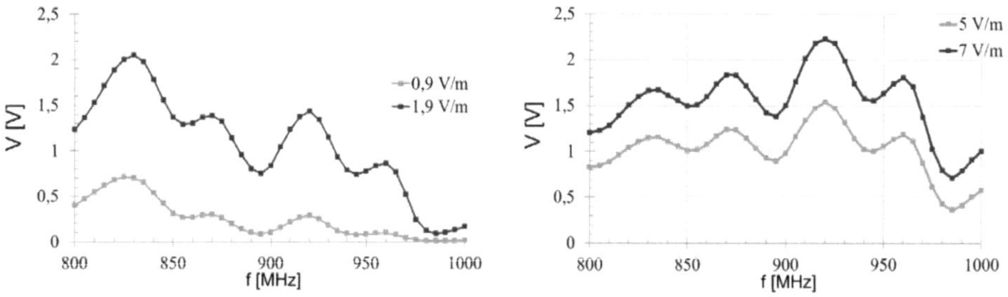

Figure 6. DC voltage on RL versus the frequency of two different values of RL: 100kΩ (left) and 10kΩ (right) and two different values of the electric field.

4. Conclusions

In this work we have presented the experimental results of an energy harvester module composed by a balanced RF-DC power converter and a matched broadband printed antenna The overall performances show that output DC voltage higher than 1V from 800MHz to 970MHz can be obtained with a load of 1kΩ.

[1] FCC 15.247. [Online]. http://www.ecfr.gov
[2] ETSI 300 220. [Online]. http://www.etsi.org/deliver/etsi_en/300200_300299/30022001/02.04.01_40/en_30022001v020401o.pdf
[3] J. Yin, J. Yi, M. K. Law, Y. Ling, M.C. Lee, K. P. Ng, B. Gao, H. C. Luong, A. Bermak, M. Chan, W.H. Ki, C.Y. Tsui, and M. Yuen, 2010 *IEEE J. Solid State Circ.*, **45** 2404.
[4] T. Le, K. Mayaram, and T. Fiez, 2008 *IEEE J. Solid State Circ.*, **43** 1287.
[5] H. Chiu, M. Lin, C. Lin, I. Ho, W. Lin, P. Fang, Y. Lee, Y. Wen, S. Lu, 2010 *IEEE Trans. Biomedical Circ. Syst.* **4** 350.
[6] Vincetti L, Maini M, Pinotti E, Larcher L, Scorcioni S, Bertacchini A,Grossi D, Tacchini A 2012, *Int. Conf. on Electromagnetics in Adv. App. -ICEAA12* (Cape Town).
[7] Scorcioni S, Larcher L, Bertacchini A, 2013 *IEEE Microwave and Wireless Comp. Letters*,.**23** 155.
[8] Dickson J.F., 1976 *IEEE Journal of Solid-State Circuits*, **11** 374.
[9] Son H.-W. and Pyo C.-S., 2005 *Electronics Letters*, **41** 994.
[10] Scorcioni S, Larcher L, Bertacchini A, 2012 *IEEE International Conference on RFID*, p 47.

IOP Conf. Series: Materials Science and Engineering **67** (2014) 012011 doi:10.1088/1757-899X/67/1/012011

Lifetime optimization of wireless sensor network by a better nodes positioning and energy distribution

J M Lebreton and N M Murad

LE2P Laboratory, 40 Avenue de Soweto, 97410 Saint-Pierre, Reunion Island, France

E-mail: jean.lebreton@co.univ-reunion.fr

Abstract. The purpose of this paper is to propose a method of energy distribution on a Wireless Sensor Network (WSN). Nodes are randomly positioned and the sink is placed at the centre of the surface. Simulations show that relay nodes around the sink are too much requested to convey data, which substantially reduces their lifetime. So, several algorithmic solutions are presented to optimize the energy distribution on each node, compared to the classical uniform energy distribution. Their performance is discussed in terms of failure rate of data transmission and network lifetime. Moreover, the total energy distributed on all nodes before the deployment is invariable and some non-uniform energy distributions are created. Finally, simulations show that every energy distributions greatly improve the WSN lifetime and decrease the failure rate of data transmission.

1. Introduction

The WSN gradually intrude into our lives and revolutionizing the way we live and work. They are found in the fields of industry, health, transportation. The sensor nodes are able to collect, process and relay a variety of data types [1]. Communication between nodes can become impossible in a certain region and this region is called a hole. Different kinds of holes are presented in [2]. In this case, the hole is caused by energy depletion among nodes around the sink. Relay nodes always do higher data transmission and thus consume more energy than the other nodes. Some papers deals with this problem and gives some possible solutions [3], [4].

This paper focuses on a way to extend the lifetime of the WSN by applying different energy distribution laws to the sensor nodes before the deployment and by proposing a way to randomly distribute nodes on the monitoring surface. The data routing protocol used are the GPSR and HEED algorithms [5], [6], [7]. Using each routing protocol, the simulations show that the energy distribution on nodes batteries and its position play a determining role in the WSN lifetime. Therefore, the failure rate of data transmission from the source node to the destination node is reduced by modifying these parameters. Our goal is to extend the network lifetime and reduce the failure rate of data transmission τ by a better nodes positioning and by selecting the better energy distribution on each node.

The paper is organized as follows: network model and routing models are presented in section 2. The Uniform Energy Distribution and the related results are discussed in section 3. The energy optimization is provided in section 4. Section 5 concludes our work.

2. Network model and routing models

2.1. Network topology model

The nodes are positioned in a random way around several rings of which the shared centre is the sink. This ring-shaped topology is most often used and well-adapted for the hierarchical WSN. The expected number of rings N_a and number of nodes N are fixed for a given WSN. Firstly, each node is placed in a random and equiprobable way on a ring. Then, the exact position of the node is determined by using a Normal distribution law $\aleph(R, \sigma)$ of which the centre is the ring radius.

2.2. Channel model and node model

A traditional energy model is used here, known as first order radio model. The energy consumption for transmitting E_{tx} is given by the equation below.

$$E_{tx}(\mathrm{d}) = \mathrm{E}_{elec} + \varepsilon_{amp} \times d^2 \tag{1}$$

with E_{elec} the energy dissipation to run the radio, ε_{amp} the energy dissipation to run transmit amplifier and d the node transmission range. Besides, the energy consumption for receiving and the energy consumption for processing are considered to be constant. For each simulation, the Telos B node parameters are used [8]. The simulations are then done with real parameters.

2.3. Routing models

One node detects an event in a random and equiprobable way. Nodes always transmit the measured information to the sink with the help of the other nodes used as relays. A transmission data cycle begins since a node detects an event and finishes when this piece of information has reached the sink, or when it is no more transmitted, otherwise the transmission can be stopped when there is a loss of radio link, if the relay node is dead or even a bad data routing. Concerning the routing strategies, two protocols have been chosen to transmit the data. The Greedy Perimeter Stateless Routing (GPSR) [5] transmit data with the help of the neighbouring nodes until reaching the sink. The Hybrid, Energy-Efficient, Distributed Clustering Approach (HEED) [6] makes up clusters within the WSN. The data detected by a node is transmitted to its Cluster Head (CH), which then transmit the data to other CHs in order to reach the sink node.

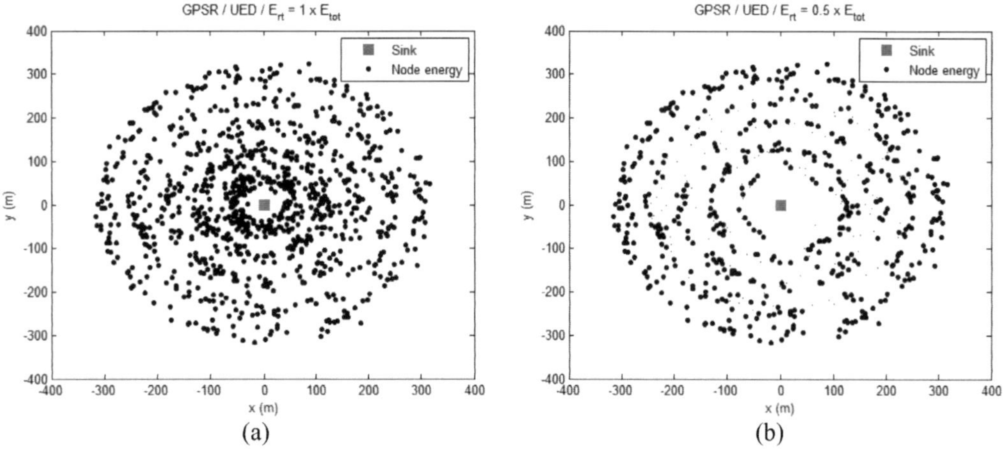

Figure 1. Evolution of the nodes remaining energy using the GPSR protocol and the UED when no energy has been consumed (a) and when 50% of the network total energy (E_{tot}) has been consumed (b). The point size is proportional to the remaining energy level in the node battery.

3. Results and discussion using the uniform energy distribution

The GPSR and HEED protocols have been used on the network and both protocols leads to an energy hole issue when the Uniform Energy Distribution (UED) is used. As illustrated in figure 1, when 50% of the network total energy has been consumed, there are few nodes with sufficient energy close to the sink. The problem is that there is still a great amount of energy available on the nodes far from the sink.

4. Energy optimization

Three energy distributions are presented here to optimize the WSN energy consumption. The main idea of these distributions is to give more energy on the nodes located close to the sink in order to avoid the energy hole issue. The simulations results are presented and discussed in section 4.2.

4.1. Energy distribution

First, the ideal case is considered in which a node would transmit directly to a node located on the adjacent ring. The Ring-shaped Energy Distribution (RED) consists of giving a different amount of energy to the nodes depending on which ring they belong. The energy proportion between the rings is calculated by considering the data transmission between nodes from one ring to another. In this scenario, 5 different levels of energy is created for the nodes.

The Optimal Energy Distribution (OEDN) is an optimal algorithm a posteriori. It distributes different amounts of energy on each node. The needed amounts of energy are computed by making several simulations of the WSN with the calculation of the nodes energy consumption.

The last distribution is a suboptimal algorithm of OEDN. The Staircase-shaped Energy Distribution (SED) calculates the average amount of required energy for different intervals of nodes distance from the sink. This distribution is quite similar to the RED if we consider the intervals like virtual rings.

4.2. Simulation results and discussion

A simulation of the identical WSN using the four energy distributions (UED, RED, NCOED and SED) has been done. Figure 2 shows the evolution of remaining total energy E_{rt} and the failure rate of data transmission τ for the GPSR and HEED protocols. The network is considered dead when τ is equal to 10% and the network lifetime is represented by the number of cycles in the graphs. It illustrates the fact that the OEDN is the most efficient distribution whatever the routing protocol. Otherwise, the RED and the SED are relatively efficient since they remarkably improve the lifetime of the WSN. In fact, the OEDN allows the network to consume almost the totality of its initial energy using the two routing protocols. This improvement is quite significant since the WSN lifetime and consumed energy are more than doubled compared to the traditional UED.

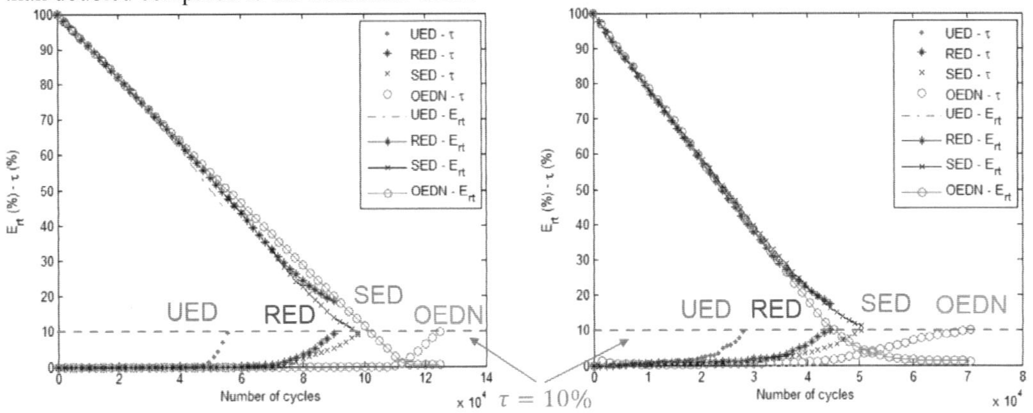

(a) (b)

Figure 2. Evolution of the remaining total energy E_{rt} and the failure rate τ using the UED, RED, SED and OEDN distributions, with the GPSR (a) and the HEED (b) protocols.

The lifetime improvement of the different energy distributions compared to the UED has been presented in the table 1. Both protocols provide similar results regarding the lifetime improvement with a particular energy distribution. Indeed, the network lifetime has been improved by about 60% using the RED and by almost 80% using the SED, still compared to the UED. These values could be considered quite low compared to the ones obtained with the OEDN, which allows the network to work more than twice longer. However, the OEDN distributes an optimal amount of energy on each node and this distribution is quite difficult to set up in a realistic scenario. In order to avoid this issue, the SED is more interesting and efficient enough to greatly improve the network lifetime. The RED also presents good results since the algorithm is simpler than the SED. Therefore, the network lifetime is significantly improved by using the new energy distributions.

Table 1. Summary of the results obtained with each distribution.

Routing protocol	Energy distribution	Lifetime improvement compared to the UED
GPSR	RED	+ 64 %
	SED	+ 78 %
	OEDN	+ 127 %
HEED	RED	+ 55 %
	SED	+ 79 %
	OEDN	+ 147 %

5. Conclusion and perspective

This article presented the energy hole issue while using a UED over the WSN. Different energy distributions have been proposed to avoid this energy hole and increase the network lifetime. The main idea of theses distributions is to distribute more energy on nodes located close to the sink. An optimal distribution has been proposed even if it is difficult to set up in a real network. Indeed, the OEDN extremely improves the reliability of the network while consuming almost the totality of the WSN energy. Besides, an a priori algorithm of energy distribution has been proposed and the RED multiplied by around 1.6 the lifetime of the WSN. Finally, the SED multiplies by 1.8 the network lifetime compared to the traditional UED. However, all these distribution algorithms prove that the WSN lifetime can be improved significantly by modifying the energy distribution.

In future works, the same study can be done to other network topologies by eventually proposing new energy distributions. Finally, the implementation of this study on a real network would be interesting too.

References

[1] Yick J, Mukherjee B and Ghosal D 2008 *Comput. Netw.* **52** 2292–330
[2] Ahmed N et al. 2005 *SIGMOBILE Mob. Comput. Commun. Rev.* **9** 4–18
[3] Wu X Chen G and Das S K 2008 *IEEE Trans. Parallel and Distrib. Syst.* **19** 710-720
[4] Szczytowski P Khelil A and Suri N 2010 *IEEE Symp. on Computers and Communications* (Riccione) (Los Alamitos: IEEE Computer Society) pp 100-106
[5] Karp B and Kung H T 2000 *Proc. of the Sixth Annual International Conf. on Mobile Computing and Networking* (Boston) (New York : ACM) pp 243-254
[6] Younis O and Fahmy S 2004 *Proc. of IEEE INFOCOM* (Hong Kong) (Los Alamitos: IEEE Computer Society) vol 1 pp 629-640
[7] Murad N 2009 *IEEE Wireless Telecommunications Symp.* (Prague, Czech Rep.) (Los Alamitos: IEEE Computer Society) pp 1-7
[8] Crossbow Technology Inc. 2013 TelosB Datasheet http://www.willow.co.uk/TelosB Datasheet.pdf

Analysis of complex environment effect on near-field emission

B Ravelo#[1], S Lallèchère*, P Bonnet* and F Paladian*
#IRSEEM EA 4353, ESIGELEC, Av. Galilée, 76801 St Etienne du Rouvray, France
*Blaise Pascal University, Pascal Institute-UMR 6602 CNRS, 24 Av. des Landais, 63171 Aubière, France

E-mail: blaise.ravelo@esigelec.fr

Abstract. The article is dealing with uncertainty analyses of radiofrequency circuits electromagnetic compatibility emission based on the near-field/near-field (NF/NF) transform combined with stochastic approach. By using 2D data corresponding to electromagnetic (EM) field (X=E or H) scanned in the observation plane placed at the position z_0 above the circuit under test (CUT), the X field map was extracted. Then, uncertainty analyses were assessed via the statistical moments from X component. In addition, stochastic collocation based was considered and calculations were applied to planar EM NF radiated by the CUTs as Wilkinson power divider and a microstrip line operating at GHz levels. After Matlab implementation, the mean and standard deviation were assessed. The present study illustrates how the variations of environmental parameters may impact EM fields. The NF uncertainty methodology can be applied to any physical parameter effects in complex environment and useful for printed circuit board (PCBs) design guideline.

1. Introduction

The electromagnetic compatibility (EMC) and complexity are major factors susceptible to limit most of electronic system performances. So, relevant EMC emission analyses [1-3] were needed to predict the sources of the harmful perturbations especially during the electronic devices design phase. However, so far, few investigations were performed on the influence of emission near-field (NF) data accuracy for EMC applications. For certain use cases, slight design imperfections against severe EMC effects may involve drastic electrical degradations. Until now, not much existing work [4] was available for the inaccuracy investigation on planar circuit EMC emissions. In this paper, we are dealing with the consideration of potential uncertainties around relative electronic boards locations. We propose herein an uncertainty analysis based on the SC model introduced in [5] applied to planar EM data for the radiated EMC emission by using NF/NF transform [6-7]. To do this, the observation plane geometrical location is assumed as deterministic unknown with an infinite precision to assess the NF scanned data sensitivity analysis.

2. Theoretical methodology on the EM NF prediction and uncertainty estimation

To measure the electric (E-) and magnetic (H-) fields emission (denoted by $X(x,y,z)$) radiated by the circuit under tests (CUTs) supposed placed in $(Oxyz)$ Cartesian system coordinate, we use the NF scanner with synoptic overview presented in Fig. 1. However, the NF scanning can take more than one hour for sweeping observation surface with $N_x \times N_y = 25 \times 40$ samplings. Therefore, we intend to use the

[1] To whom any correspondence should be addressed.

NF/NF transform $X(z)=f[X(z_0)]$ developed in [6-7] to extract the EM data $X(z)$ at any position z from the given data $X(z_0)$ in the plane placed at $z=z_0$ in both the frequency- and time-domains.

2.1. Preliminary investigation on EMC NF emission

The NF/NF method exploited in this paper is based on the plane wave spectrum (PWS) theory [3]. The error vector magnitude (EVM) relative corresponding to an empirical example of this NF/NF transform are plotted in Fig. 2 for a CUT radiation in the observation plane placed from 2mm to 10mm. The EMC sensitivity analysis linked to this EVM was evaluated by using a stochastic model in function of the geometrical position whose mechanism and methodology are depicted in Fig. 3.

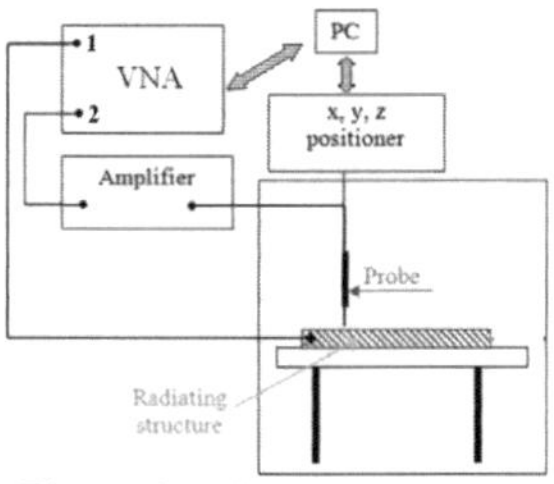

Figure 1. *IRSEEM NF scanner illustrative diagram.*

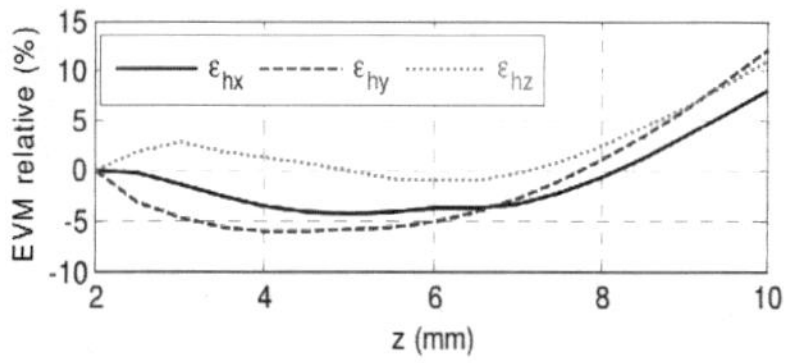

Figure 2. *Relative EVM from NF/NF transform.*

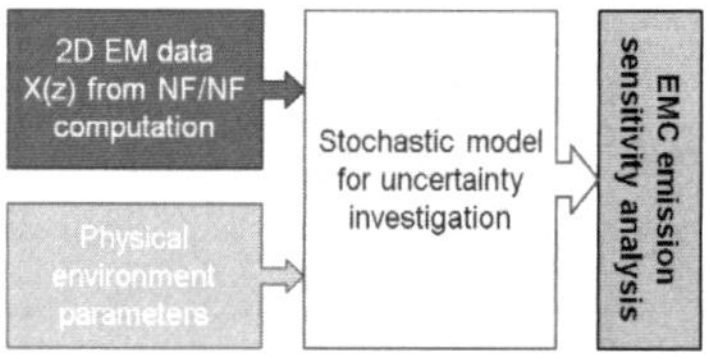

Figure 3. *Methodology for investigating EMC emission sensitivity analysis.*

2.2. Mathematical approach for the uncertainty analysis under consideration

The proposed uncertainty analysis is fundamentally established with a polynomial approximation of given mapping data expressed by the variable F which is supposed depending on a random parameter $Z=Z^0+u$ (Z^0 as initial value and u as random variable (RV)). Firstly, we proceed with a n-order Lagrangian expansion of function $S \rightarrow F(Z^0,S)$:

$$F(Z^0,S) \approx \sum_{k=0}^{n} F_k(Z^0)\cdot L_k(S)\,, \qquad (1)$$

with L_k (k is a positive integer) is the Lagrangian polynomial and requiring $(n+1)$ SC points. Based on the demonstration in [5], this expansion can be defined with weighted coefficient ω_k, set of points S_m. One of the most interesting properties of the Lagrangian basis relies on its reducing characteristic: $L_k(S_m)=\delta_{km}$ (Kronecker *symbol*) which justifies the parameter choice by greatly simplifying weights computing [5]. Hence, for instance, the mean and variance of mapping F are written as follows:

$$\langle F(Z^0,S)\rangle \approx \sum_{k=0}^{n} \omega_k \cdot F_k(Z^0) \qquad (2) \qquad \mathrm{var}(F(Z^0,S)) \approx \sum_{k=0}^{n} \omega_k^2 \cdot F_k^2(Z^0) - \langle F(Z^0,S)\rangle^2 \qquad (3)$$

Table 1 addresses the calculated weights and collocation point coordinates for $n=\{2,3,4\}$.

Table 1. SC weighted points with random variation along z axis: $z=z^0+u$.

SC order n	Weights (no dimension)	Points (in mm)
2	(0.277, 0.444, 0.277)	(6.45, 8.00, 9.55)
3	(0.174, 0.326, 0.326, 0.174)	(6.28, 7.32, 8.68, 9.72)
4	(0.118, 0.239, 0.284, 0.239, 0.118)	(6.19, 6.92, 8.00, 9.08, 9.81)

As figure of merit, the coefficient of variation (CV) is defined as the ratio between standard (std) and mean of EM field components. This approach relevance was checked with Matlab and use cases of EM scanned and simulated radiated data versus the observation plane positions.

3. Uncertainty analysis applications with experimental and simulation data

The Wilkinson power divider (WPD) operating at 1GHz and the arbitrary shaped microstrip line whose photo and HFSS 3D design are displayed, respectively, in Figs. 4(a) and 4(b) are considered as planar CUT1 and CUT2. The IRSEEM NF scanner monitored in Fig. 5 was used to generate the measured scanned EM NF input data exploited in this investigation. It is worth noting that the NF/NF transform was validated experimentally and by EM computations with various structures [6].

Figure 4(a). *WPD.*

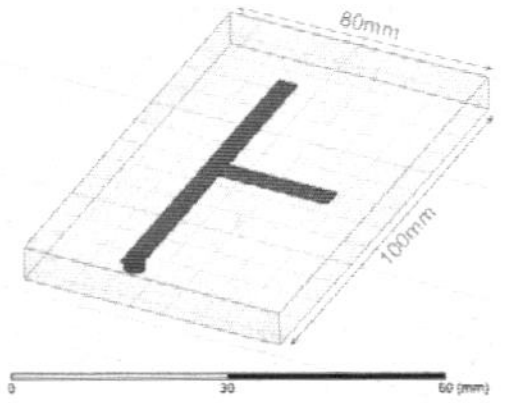

Figure 4(b). *Stub microstrip line.*

Figure 5. *IRSEEM NF scanner.*

3.1. Uncertainty Analysis of NF emission from the WPD

The numerical codes of the methodology shown in Fig. 3 was implemented and executed to provide a huge deterministic database with 6501 samples between z_0=2mm and 15mm, step Δz=2μm. Based upon computed NF data and in order to validate SC accuracy, we realised classical Monte Carlo (MC) simulations. By considering the stochastic modelling of random parameter z addressed in Table 1, we generate the first and second statistical moments of E-field components (same results without any additional cost for H-field) in Fig. 6 for z_0=2mm @ 1GHz.

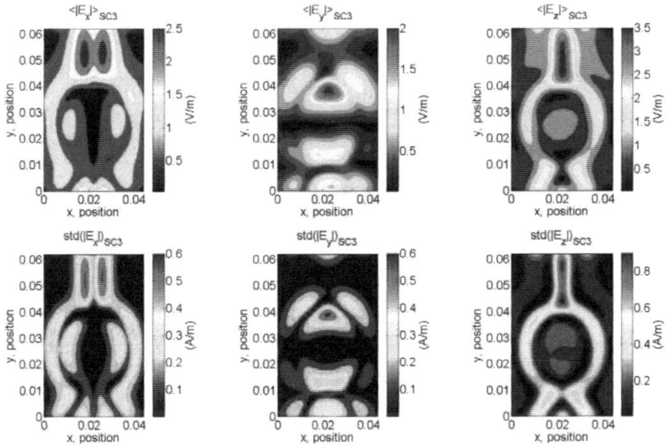

Figure 6. *WPD (Fig. 4(a)) NF scanning: E-field mean (in top) and std (in bottom) from SC (n=2, 3 points, SC3) @ 1GHz.*

Table 2 summarizes the overview both of the convergence requirements (MC, SC) and SC precision considering relative differences of space averaged results over the entire scanning plan. In accordance with the relative gap existing between different MC realizations and various SC orders, the differences between means of E-fields are about 5% by considering 10MC and 100MC simulations; they decrease of about 1.4% averaging the results from 1000MC realizations. The good agreement between SC results (only 3 simulations needed with some ms computation time) and MC with 1000 realizations (relative gap around 0.05%) demonstrates the SC accuracy and efficiency.

Table 2. Relative statistical (spatial mean value) gaps over entire E_x and H_z cartographies.

Statistics from E-fields	MC10/100	MC100 /1000	SC3/MC1000	SC3/SC5
mean(E_x)	5.71%	1.37%	0.05%	0.024%
std(E_x)	5.20%	7.75%	0.67%	0.27%

The SC convergence of about 0.02% is obtained by comparing data from SC orders n=2 and n=4 (SC3/SC5). Same remarks can be found from results for E_x and H_z std deviations, validating SC convergence for second statistical moment computing. Consequently, the accuracy results from Fig. 6 provide the system sensitivity, setting the focus on crucial locations. As expected, due to WPD radiation cartographies under consideration, most sensitive areas are centred around the WPD position.

3.2. Sensitivity analysis of microstrip line NF emissions

In this example, different cases showing how the SC may provide additional information are analysed. A set of deterministic results were computed over z=5mm to 20mm, Δz=2µm @ 3GHz. Without any lack of generality and in order to demonstrate the SC technique robustness, the uncertainty was modeled assuming a uniform distribution in Table 1. The SC precision remains very close to thousands of MC simulations with high convergence rate (relative gap between SC3 and SC5 around 1.5% depending on the EM components) and only 3 simulations are required to compute the mean value, std and CV. Fig. 7 provides the std and CV results for H-field, and lay emphasis on the most sensitive locations (due to random z-variations) with a quantitative assessment of expected effects. We can see that based on the cartography of the most sensitive areas which limits computing efforts, the H-field dispersion is around the mean value varies between 0.05A/m and 0.16A/m.

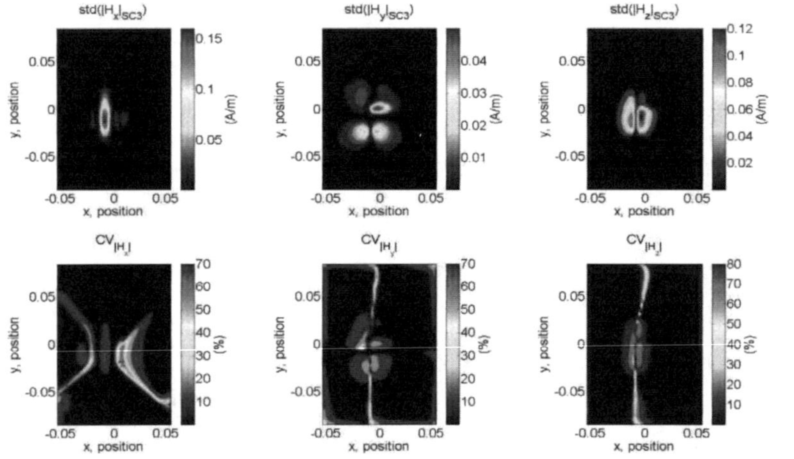

Figure 7. Microstrip line (Fig. 4(b)) NF radiation analysis: H-field std (in top) and CV (in bottom) from SC3 @ 3GHz.

4. Concluding remarks

A computational method of EMC emission uncertainty analyses established from NF/NF transform is investigated. The analytical approach illustrating the uncertainty model functioning principle is introduced based on the SC approach introduced in [6]. As application, non-intrusive, quick, precise and relevant sensitivity analysis of EM NF data radiated by microstrip circuits as WPD and interconnect line versus position are provided. This uncertainty analysis method is useful to check the geometrical inaccuracies influence on radiated EMC emissions during the design cycle.

5. Acknowledgements

This research work is co-funded by the Upper-Normandy Region and the ERDF within the frame of the "PULser electro optique pour Sonde Atomique Tomographique (PULSAT)".

6. References

[1] Ostermann T and Deutschmann B 2003 *Proc 13th ACM Great Lakes Symp. VLSI (Washington, DC, 28-29 Apr 2003)* pp 76-79
[2] Gonser M, Keller C, Hansen J, Khillkevich V, Radchenko A, Pommerenke D and Weigel R 2012 *IEEE MTT-S IMS Digest* (*Montreal, QC, 17-22 June 2012*) pp 1-3
[3] Haixiao W, Beetner DG and DuBroff RE 2011 *IEEE Tran EMC* **53** pp 891-899
[4] Arnaut L, Obiekezie CS, and Thomas DWP 2013 *IEEE Tran EMC* **99** pp 1-11
[5] Paladian F, Bonnet P and Lalléchère S 2011 *Proc XXXth URSI GASS* Istanbul Turkey
[6] Leseigneur C, Baudry D, Ravelo B and Louis A 2013 *EPJAP* **64** 11001 pp 1-14
[7] Liu Y and Ravelo B 2012 *Advanced Electromagnetics* **1** pp 24-40

Hybrid modelling of near-field coupling onto grounded wire under ultra-short duration perturbation

B Ravelo[1], **Y. Liu**
IRSEEM EA 4353, ESIGELEC, Av. Galilée, 76801 St Etienne du Rouvray, France

E-mail: blaise.ravelo@esigelec.fr

Abstract. A time-frequency (TF) hybrid model (HM) for investigating the interaction between EM near-field (NF) aggression and grounded wire is addressed. The HM is based on the combination of techniques for extracting the EM NF radiated by electronic structures and the calculation of electrical disturbances across the wire due to EM coupling. The computation method is fundamentally inspired from transmission line (TL) theory under EM illumination. The methodology including flow chart interpreting the routine algorithm based on the combination of frequency and time domain approaches is featured. An experimental result showing the EM coupling between patch antenna-wire from 1.5-3.5GHz reveals the efficiency of the HM in frequency domain. The relevance of this HM was illustrated with a structure comprised of 20cm aggressor and 5cm victim I-shaped wires placed above a planar ground plane. The aggressor was excited with 40ns duration perturbation signal. After Matlab implementation of the HM, the disturbance voltages across the extremity of the victim wire were extracted. This simple and fast HM is useful for the EMC engineering during the design and fabrication phases of electrical and electronic systems.

1. Introduction

The modern electronic systems integration density and complexity are unintentionally threatened by electromagnetic compatibility (EMC) problems. The electrical wires interconnecting the electronic systems are usually involved. All level of electronic engineering as automotive and aeronautic and so forth is concerned by the occurrence of harmful EM coupling between nearby electronic component radiation causing embarrassing malfunctioning [1-3]. As the EMC analysis of these structures is challenging, methods enabling to predict disturbance voltages and currents, for example, across the victim transmission line (TL) under EM illumination were introduced [4-7]. At the beginning, the both E- and H-fields effects on the TL were proposed in [4], then it was adapted to only E- [5] and H- [6] fields. Those methods were limited to the TL interacting with planar and uniform EM far-fields [4-6]. More recently, the computational method was extended to cases of more complex systems composed of radiating circuits but operating only in a very limited band frequency [7]. It means that a more efficient approach is needed to extend this method for the case of radiations in wide frequency range as the case of surge currents, burst, lightning and also standard transient signals as identified in IEC/EN 61000-4-2. The aim of this paper is to propose a simple and fast time-frequency (TF) model of the EM near-field (NF) coupling onto grounded wires by taking into account ultra-short duration perturbations as electric static discharge (ESD) and electrical overstress phenomena.

[1] To whom any correspondence should be addressed.

2. Methodology of the time-frequency (TF) hybrid method (HM) for extracting the disturbance from EM NF coupling

A radiating electronic device placed in proximity of planar grounded wire constitutes the structure understudy [7]. The proposed TF HM consists in characterizing the transient EM radiation and the disturbance across the wire by coupling voltages. These couplings are generated in function of the perturbation signals on the aggressor circuit convoluted with coupling transfer function.

Fig. 1 explains the functioning principle and indicates the five main steps of the proposed model. ***i) Step 1*** is aimed to the determination of EM data radiated by the aggressor. This can be obtained with different ways, from measurements [1] or simulations with the 3D EM tools or models [2-3] or both combined [7]. The NF data provided is generally expressed in the frequency domain (FD) as the case of standard IEC61967. ***ii) Step 2*** is the analytical modelling of the victim under the EM aggression according to TL equations. ***iii) Step 3*** is the extraction of the HM transfer function, $T_x(j\omega)=U_x(j\omega)/U(j\omega)$ ($j\omega$ is the radian frequency variable) between the excitation signal $U(j\omega)$ and the disturbance $U_x(j\omega)$ induced at the extremity of the victim line. ***iv) Step 4*** is the spectral analysis of the perturbation signal exciting the aggressor. Then, ***v) Step 5*** is the convolution between the perturbation signal and the HM transfer function to determine the disturbance in the time-domain (TD) via IFFT.

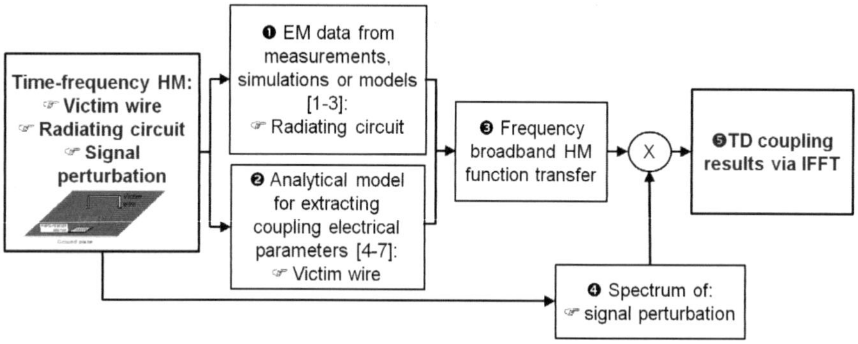

Figure 1. Flow chart illustrating the methodology of the proposed TF HM.

For the theoretic illustration, the victim wire loaded by $Z(x=0)$ and $Z(x=d)$ is with length d, placed at the height z_0 above the ground plane in ($Oxyz$) Cartesian system and oriented along x-axis as in [5-7]. The circuit model of this wire is formed by LC-network characterized by per-unit length inductance L_u and capacitance C_u. With this circuit model, we use the Agrawal model [5] to determine the disturbance or coupling voltages $U(x=0)$ and $U(x=d)$ thanks to its simplicity. In this case, the coupling equations require only E_x component as EM input data. More concretely, by denoting ω the radian frequency, the differential system generating the wire disturbance can be materialized by:

$$\partial U(x)/\partial x + j\omega \cdot L_u \cdot I(x) = E_x(x, z = z_0),\ \partial I(x)/\partial x + j\omega \cdot C_u \cdot U(x) = 0, \quad (1)$$

with the boundary conditions: $U(x=0)=Z(x=0)I(x=0)$ and $U(x=d)=Z(x=d)I(x=d)$.

3. Applications

3.1. Experimental validation in the FD

Fig. 2(a) represents the experimental setup and layout of the device under test (DUT). It consists of a patch antenna resonating at about 3GHz whose layout displayed in Fig. 2(b) as aggressor circuit. It is placed in proximity of d=10cm length cylindrical copper wire fixed at the heights z_1=1cm and z_2=3cm above the planar ground plane. After measurement scanning of the active electrical field component $E_x(f)$ from 1.5GHz-3.5GHz in the surface area delimited by the wire and the ground plane, we extracted the coupled voltages $U(x=0,z)=U(x=d,z)$ (due to the symmetry). In Fig. 2(c), it is displayed the comparison between the computed results from the HM and full wave simulations from HFSS commercial tool. We found a correlation between the behaviour of the coupling level which increases with the frequency despite the slight difference around the resonance frequency. The computation run time with PC equipped by operating system windows 7 having Intel® CPU @ 2.3GHz - 2Go RAM was 5 seconds compared to the HFSS simulations which took about 40 minutes.

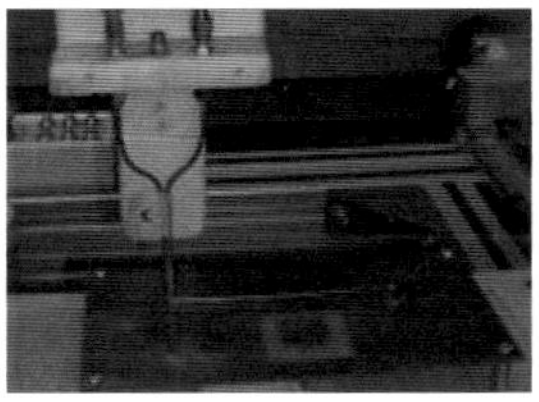

Figure 2(a). Experimental setup.

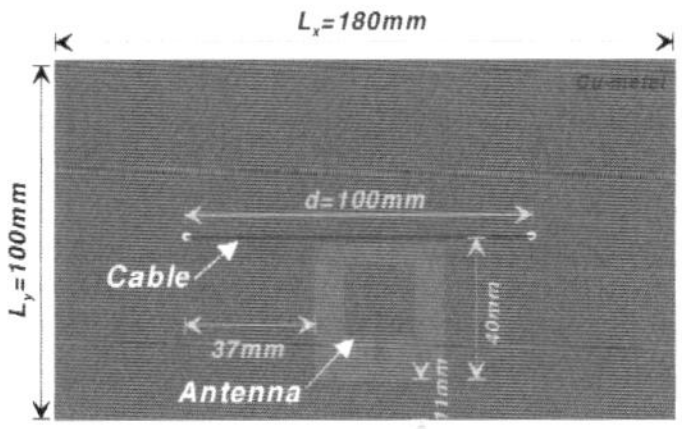

Figure 2(b). DUT layout.

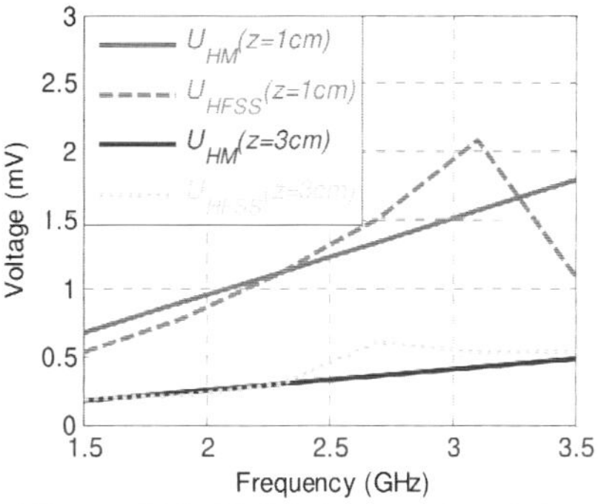

Figure 2(c). Validation results of the HM in FD.

3.2. Analysis of the TF HM with perpendicular configuration wire placed above ground planes

More general analysis including Matlab numerical applications in both the FD and TD is proposed in this subsection by considering a surge signal with 40ns duration as signal perturbation.

3.2.1. Description of the structure under study. Fig. 3 represents a structure composed of two perpendicular Copper wires (P_1P_2) and (P_3P_4). The first one is assigned with 20cm length and diameter Ø=1mm excited by a transient signal at P_1 is the aggressor. The second one (P_3P_4) oriented along *x*-axis having 5cm length and Ø=1mm and 50Ω loaded at P_3 and P_4 is set as the victim.

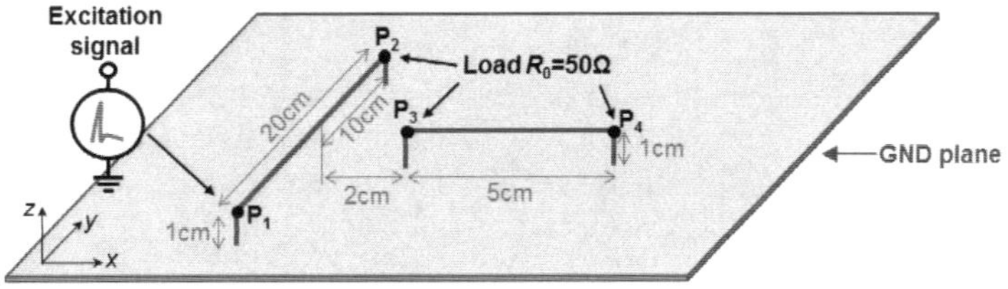

Figure 3. Configuration of the structure understudy for the TF HM analysis.

3.2.2. Computed results from the TF HM. First, we computed the TD EM radiation in the surface plane delimited by (P_3P_4) and the ground plane when the aggressor (P_1P_2) is excited by the transient signal *u*(*t*) with 40ns duration and 5V amplitude double exponential waveform (see Fig. 5(c)). Second, the EM NF emission in TD is extracted with full wave simulations, the active electric field $E_x(t)$ from the aggressor at the arbitrary instant time t_0=9ns is displayed in Fig. 4(a). Then, via FFT, we generated the equivalent data $E_x(f)$ in the FD from DC to about 0.5GHz.

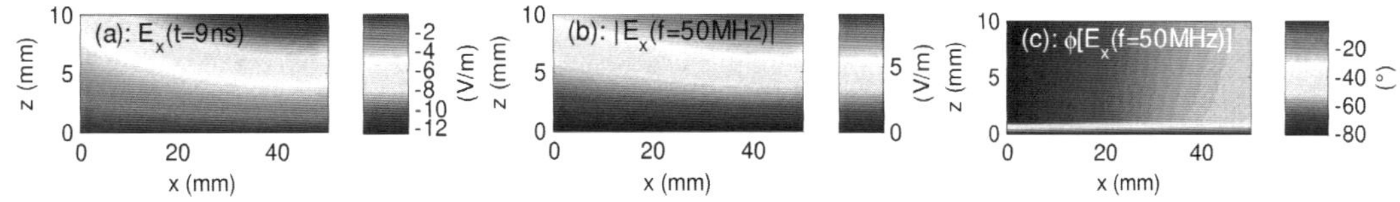

Figure 4. Maps of the E-field active component: (a) TD real component, and FD (b) magnitude and (c) phase in the surface plane delimited by the victim wire and GND plane.

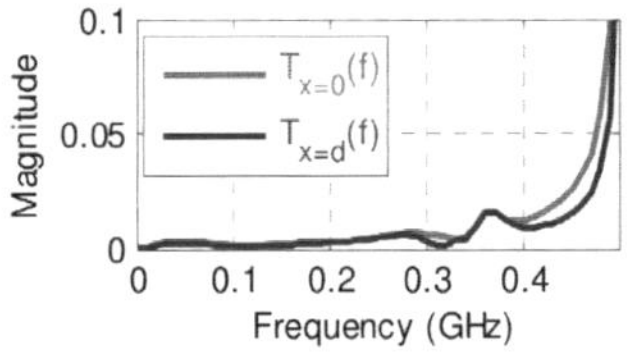

Figure 5(a). Coupling transfer function magnitude.

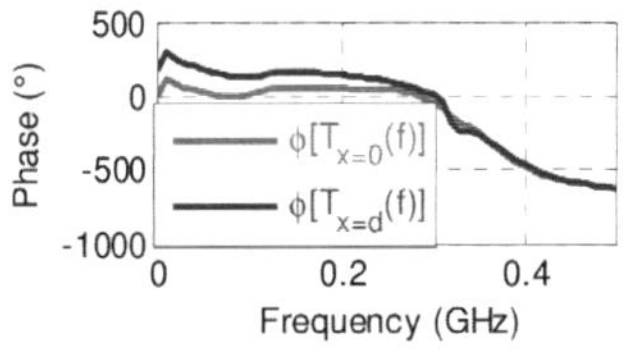

Figure 5(b). Coupling transfer function phase shift.

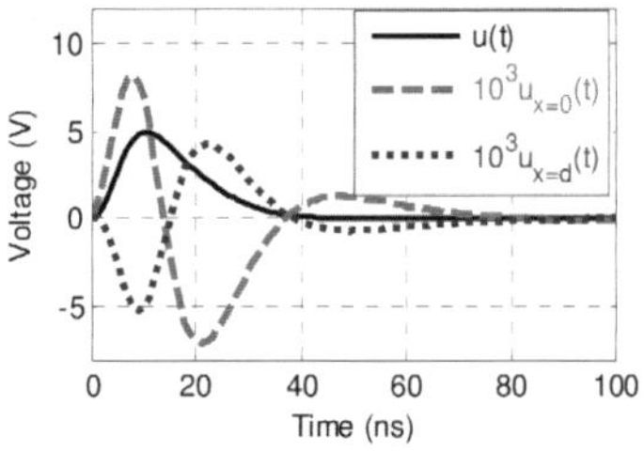

Figure 5(c). TD voltage coupling disturbances.

The E-field amplitude and phase maps at the arbitrary frequency f_0=50MHz are shown in Figs. 4(b) and 4(c). To determine the couplings, the equivalent circuit model of the victim line should be drawn. The LC-model of (P_3P_4) presents per unit length parameters L_u=518nH/m, C_u=21.5pF/m and characteristic impedance Z_c=155Ω. We point out the victim wire conductive loss was neglected. Thanks to the methodology instructed earlier in Fig. 1, we obtain the transfer functions $T_{x=0}(j\omega)$ and $T_{x=d}(j\omega)$ of the structure transfer functions plotted in Figs. 5(a) and 5(b). Then, we extracted the coupling voltages $U_{x=0}(j\omega)$ and $U_{x=d}(j\omega)$ in FD. Finally, via another IFFT, we calculated the time-dependent induced couplings $u_{x=0}(t)$ and $u_{x=d}(t)$, plotted in dashed lines of Fig. 5(c). As expected, we can see that the transient disturbances with mV level were found at the extremities of (P_3P_4).

4. Concluding remarks

A computational HM aimed on the investigation on the coupling between an electronic devices and planar grounded wires applied to the radiated EMC analysis is introduced and experimentally validated. The HM principle is managed by combining the EM data representing the of the aggressor radiation onto the victim wire and the TL theoretical approach for calculating the disturbances in both FD and TD. The effectiveness of the proposed modelling approach was verified in the FD with a structure composed of a patch antenna and grounded metallic wire. The results were validated with HFSS simulations. Then, another application based on the coupling between wires under transient perturbation was analysed. After determination of coupling transfer function, the transient disturbance voltages across the victim line in function of the excitation voltages were generated. Compared to the existing EMC analysis and simulation tools, the TF HM offers benefits with outstanding speed computation time and in term of flexibility to operate in both the FD and TD.

As ongoing research, the extension of this TF HM for more complex systems constituted by series of PCBs placed in harsh environment by taking into account temperature effect is one of the main ultimate targets of this work.

Acknowledgements

This research work is co-funded by the Upper-Normandy Region and the ERDF within the frame of the "PULser electro optique pour Sonde Atomique Tomographique (PULSAT)".

References

[1] Kaires R G 1998 *Proc IEEE Int Symp on EMC (Denver, CO, 24-28 Aug 1998)* pp 872-877
[2] Ravelo B, Liu Y and Slama J B H 2011 *Eur Phys J Appl Phys* **53** 30701 pp 1-8
[3] Liu Y, Ravelo B, Fernandez-Lopez P, Baudry D and Ben Hadj Slama J 2011 *Proc 5th Int Conf on EM Near-Field Characterization and Imaging (Rouen, France, 30 Nov-2 Dec 2011)* pp 1-6
[4] Taylor C D and Sattewhite R S 1965 *IEEE Tran Ant Prop* **E** pp 987-989
[5] Agrawal A K and Price H J 1980 *IEEE Tran Ant Prop* **18** pp 432-435
[6] Rachidi F 1993 *IEEE Tran EMC* **35** pp 404-407
[7] Leseigneur C, Baudry D, Ravelo B and Louis A 2013 *Eur Phys J Appl Phys* **64** 11001 pp 1-14

IOP Conf. Series: Materials Science and Engineering **67** (2014) 012014 doi:10.1088/1757-899X/67/1/012014

Transient Response of Grounding Electrode using the Wire Antenna Theory Approach

D. Poljak[1], S. Sesnic[1], K. El Khamlichi Drissi[2] and K. Kerroum[2]
[1]University of Split, FESB, Split, Croatia
[2]Blaise Pascal University, Pascal Institute, Clermont-Ferrand, France
E-mail: dragan.poljak@fesb.hr

Abstract. The paper deals with an assessment of transient current induced along the vertical grounding electrode applying the thin wire antenna model in the frequency domain. The formulation is based on the homogeneous Pocklington integro-differential equation. The influence of an imperfectly conducting half-space is taken into account via the corresponding reflection coefficient arising from the Modified Image Theory (MIT). The Pocklington equation is solved using numerical and analytical approach respectively. The numerical solution is carried out via the Galerkin-Bubnov scheme of the Indirect Boundary Element Method (GB-IBEM). The transient response is obtained by means of the Inverse Fourier Transform (IFFT). Some illustrative computational examples for the transient current induced at the center of a vertical grounding electrode are presented in the paper.

1. Introduction

Transient analysis of vertical grounding electrode, as an important component in many realistic grounding systems, such as wind turbines (WTs) is always of interest in electromagnetic compatibility (EMC) [1], [2]. Transient analysis of complex grounding systems for wind turbines has been already reported elsewhere, e.g. in [1], where a combination of the ring, vertical rods and horizontal electrodes is considered. Nevertheless, a study of vertical electrode behavior, as an individual element, still remains a rather important issue.

Transient behavior of wire structures in EMC can be carried out using the approximate Transmission Line (TL) approach, or using the more rigorous wire antenna theory (AT) model [3], [4]. A trade-off between AT and TL approach can be found elsewhere, e.g. in [4]. Frequency domain numerical/analytical antenna theory study of vertical electrodes as individual elements, using Galerkin-Bubnov scheme of the Indirect Boundary Element Method (GB-IBEM) and analytical solution, respectively, has been reported in [2].

This paper deals with a transient response of a straight vertical electrode, thus extending the frequency domain analysis presented in [2]. First, the current distribution along the electrode is obtained by using both the GB-IBEM and analytical solution procedure presented in [2].

The transient response of the vertical electrode is obtained by means of Inverse Fourier Transform (IFFT). Some illustrative computational results are given in the paper.

2. Formulation

Fig. 1 shows a vertical grounding electrode of length L and radius a, buried in a lossy medium at depth d and energized by an equivalent current generator with one terminal connected to the electrode and the other one grounded at infinity.

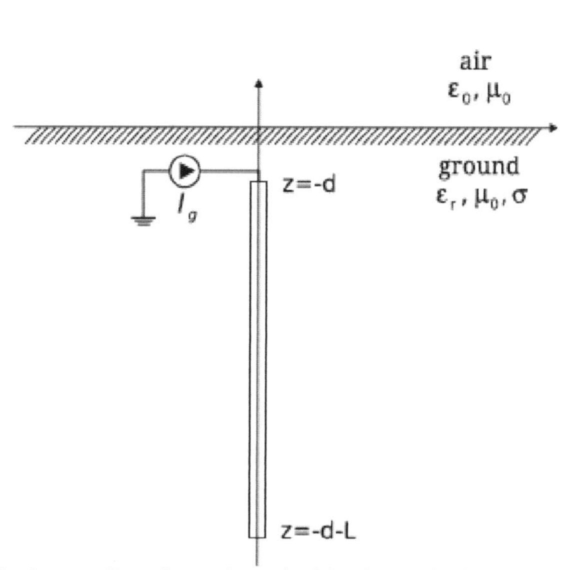

Fig. 1. Vertical grounding electrode excited by the equivalent current generator.

The governing integral expressions for the current and scattered voltage induced along the vertical electrode can be readily derived from Maxwell's equations by enforcing the continuity conditions for the tangential components of the electric field along the electrode and utilizing the concept of generalized Telegrapher's equations.

The current distribution along the grounding electrode is governed by a homogeneous integro-differential equation of the Pocklington type [2]:

$$j\omega\frac{\mu}{4\pi}\int_{-d-L}^{-d} I(z')g(z,z')dz' - \frac{1}{j4\pi\omega\varepsilon_{eff}}\frac{\partial}{\partial z}\int_{-d-L}^{-d}\frac{\partial I(z')}{\partial z'}g(z,z')dz' + Z_s(z)I(z) = 0. \tag{1}$$

The complex permittivity of the lossy ground ε_{eff} is

$$\varepsilon_{eff} = \varepsilon_r\varepsilon_0 + \frac{\sigma}{j\omega}, \tag{2}$$

where ε_r and σ are the soil permittivity and conductivity, respectively.

The total Green function is of the form [2]:

$$g(z,z') = g_0(z,z') - \Gamma_{ref} g_i(z,z'), \tag{3}$$

where $g_0(z,z')$ and $g_i(z,z')$ are given by:

$$g_0(z,z') = \frac{e^{-\gamma R_1}}{R_1}, \quad g_i(z,z') = \frac{e^{-\gamma R_2}}{R_2} \tag{4}$$

and the propagation constant of the lossy ground is:

$$\gamma = \sqrt{j\omega\mu\sigma - \omega^2\mu\varepsilon_r\varepsilon_0}, \tag{5}$$

where R_1 and R_2 are the distances from the source and the image to the observation point, respectively.

Within the numerical solution the Fresnel reflection coefficient (RC) is used [4]:

$$\Gamma_{ref}^{MIT} = \frac{\frac{1}{n}\cos\theta - \sqrt{\frac{1}{n} - \sin^2\theta}}{\frac{1}{n}\cos\theta + \sqrt{\frac{1}{n} - \sin^2\theta}};\ \theta = arctg\frac{|x-x'|}{2d};\ \underline{n} = \frac{\varepsilon_{eff}}{\varepsilon_0}. \tag{6}$$

Furthermore, a simplified reflection coefficient arising from modified image theory is used within the analytical solution and is given by [4]:

$$\Gamma_{ref}^{MIT} = -\frac{\varepsilon_{eff} - \varepsilon_0}{\varepsilon_{eff} + \varepsilon_0}. \tag{7}$$

The current source is inserted into the Pocklington integro-differential equation formulation in terms of the following conditions at the electrode ends [2]:

$$I(-d) = I_g,\ I(-d-L) = 0, \tag{8}$$

where I_g stands for the unit current source excitation.

It is worth noting that the knowledge of the electrode current provides the assessment of the scattered voltage.

The scattered voltage along the electrode can be obtained by integrating the horizontal component of the electric field (normal to the electrode and tangential to the ground-air interface) from the remote soil to the electrode surface [5].

Performing certain mathematical manipulations and assuming the scalar potential in the remote soil to be zero [5], yields:

$$V^{sct}(z) = -\int_{\infty}^{a} E_x^{sct}(x,z)dx = -\frac{1}{j4\pi\omega\varepsilon_{eff}}\int_{-d-L}^{-d}\frac{\partial I(z')}{\partial z'}g(z,z')dz' \tag{9}$$

Note that the tedious integration from infinity to the electrode surface is avoided in the final expression at the right hand side. The procedure is presented in detail in [5].

3. Solution procedures

The space-time dependent current and voltage along the electrode are obtained by solving (1) and (9) using numerical/analytical solution procedures. The full mathematical description of the solution procedures can be found elsewhere, e.g. in [2] and [4].

The transient response of the vertical electrode is calculated by using the IFFT procedure [6].

Thus the discrete frequency response is obtained by sampling analytical solution in the frequency domain:

$$I(x,\omega_i) = \sum_{i=1}^{N} I(x,\omega)\delta(\omega - \omega_i) \tag{10}$$

Transient current induced at the center of the electrode is calculated using the following IFFT formula:

$$I(x,t_j) = \frac{1}{N}\sum_{i=1}^{N} I(x,\omega_i)\omega_N^{-(j-1)(i-1)} \tag{11}$$

Having performed extensive numerical testing, 2^{16} samples with maximum frequency of 500 MHz were found to be optimal to ensure accurate results within a reasonable time frame.

4. Results

Some illustrative computational examples for the transient current induced at the center of 1 m and 10 m long electrode obtained by means of analytical and numerical approach, respectively are presented in Figs. 2 and 3. Radius of the electrode is a=5 mm, ground relative permittivity is, ε_r=10, burial depth is d=0.5 m. The electrode is excited by the double-exponential 0.1/1 μs pulse.

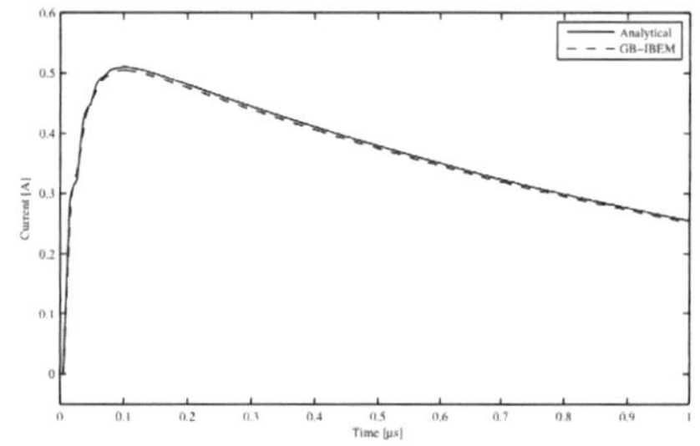

Fig. 2. Transient current at the center of the grounding electrode, L=1 m, σ=10 mS/m.

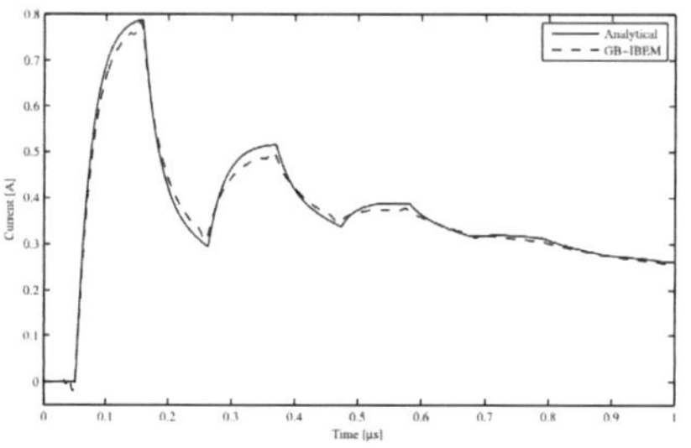

Fig. 3. Transient current at the center of the grounding electrode, L=10 m, σ=1 mS/m.

Note that in both cases the results obtained via different approaches agree quite satisfactorily.

5. Conclusion

Transient behavior of the vertical grounding electrode has been analyzed using the thin wire antenna model. The formulation is based on the homogeneous Pocklington equation in the frequency domain. The influence of a dissipative half-space is taken into account via the corresponding reflection coefficient. The Pocklington equation is solved using numerical and analytical approach respectively. The numerical solution is carried out via the Galerkin-Bubnov variant of the Indirect Boundary Element Method (GB-IBEM). The transient response is calculated by means of the Inverse Fourier Transform (IFFT). Some illustrative computational examples for the transient current induced at the center of 1 m and 10 m long vertical grounding electrode are presented in this work. The results obtained via different approaches are found to be in a satisfactory agreement.

References

[1] D. Cavka, D. Poljak, V. Doric, R. Goic, Transient Analysis of Wind Turbines, Renewable energy 43, pp. 284-291, 2012.

[2] D. Poljak, S. Sesnic, Wire antenna Model of the Vertical Grounding Electrode, *Boundary Elements and Other Mesh Reduction Methods XXXV*, BEM 2013, New Forest, UK, Sept. 2013, pp. 103-112.

[3] F. M. Tesche, M. Ianoz, and T. Karlsson, EMC Analysis Methods and Computational Models. New York, USA: John Wiley & Sons, Inc., 1997.

[4] D. Poljak, Advanced Modeling in Computational Electromagnetic Compatibility. New Jersey, USA: Wiley-Interscience, 2007.

[5] D. Poljak, V. Doric, F. Rachidi, K. El Khamlichi Drissi, K. Kerroum, S. Tkachenko, S. Sesnic, Generalized Form of Telegrapher' s Equations for the Electromagnetic Field Coupling to Buried Wires of Finite Length, *IEEE Trans. on Electromagnetic Compatibility*. 51 (2009), 2; 331-337.

[6] S. Šesnić and D. Poljak, "Frequency and Time Domain Analysis of Influence of the Grounding Electrode Conductivity on Induced Current Distribution," *Journal of Communications Software and Systems*, vol. 9, no. 2, pp. 137-143, 2013.

Electromagnetic game modeling through Tensor Analysis of Networks and Game Theory

Olivier Maurice, Alain Reineix, Sébastien Lalléchère
GERAC, 3 av. Jean d Alembert, 78190 Trappes, FRANCE
E-mail: mauriceo@gerac.com

Abstract. A complex system involves events coming from natural behaviors. Whatever is the complicated face of machines, they are still far from the complexity of natural systems. Currently, economy is one of the rare science trying to find out some ways to model human behavior. These attempts involve game theory and psychology. Our purpose is to develop a formalism able to take in charge both game and hardware modeling. We first present the Tensorial Analysis of Networks, used for the material part of the system. Then, we detail the mathematical objects defined in order to describe the evolution of the system and its gaming side. To illustrate the discussion we consider the case of a drone whose electronic can be disturbed by a radar field, but this drone must fly as near as possible close to this radar.

1. Introduction

Complex problems cannot be approached through classical techniques. One reason comes from the diversity of elements that compose this complexity. Deterministic tools can be used once a first understanding of the problems is covered by analysis and physical looking. Moreover, psychological side of complex systems like human ones must incorporate notions of choices, payoff, etc.; these notions that doesn't belong to numerical tools. Anyway, reality has to be symbolized to dispose of some representation of the global problem. In order to realize this task, networks give an efficient tool. Once can refer to numerous works on heat science, mechanics, computational fluid dynamics, biochemistry [1] ..., and of course in electricity where they are usually employed. These networks are made of nodes and edges, they can be extended using meshes and chords. The chords allow to support any kind of mathematical interaction between two nodes, edges or two meshes. We propose here for example using chord to take in charge the electromagnetic interaction between the radar and a moving target.

2. Physical modeling

As explained in the introduction, a global problem can be decomposed into two part, the first one that consists in the modeling of the physical problem using the network theory. The systems under study consists in a fixed radar and a moving target. Each previous subsystem and their interactions should be modeled. Sir G. Kron was the first to understand the benefit of using tensorial algebra in such networks study [2]. As the systems may change over time, an extension of the method has been proposed by the authors [3]: it consists in the definition of tenfolds as explained later that will be able to follow an evolution using gamma matrix [4].

2.1. Physical issue: the Kron's method

The system under study includes two main objects: the radar and the target. Each one can be represented by a so called primitive network under the Tensorial Analysis of Network (TAN) formalism so they can exist independently of the system. To create our scene the operation consists in coupling the radar and the target through an electromagnetic interaction, this being made on a operational theater. So the system modeling is realized in three parts.

First, the radar consists in an amplifier, source of the electromagnetic energy and an antenna. Each element can be associated with an edge: one for the antenna and the other for the amplifier. Each edge is associated respectively to an impedance: Z_1 and Z_2. The models construction requires various operations. The entire system impedance is obtained from a direct summation of the objects: $g_1 = \begin{bmatrix} Z_1 & 0 \\ 0 & Z_2 \end{bmatrix}$. Next step consists in connecting the two edges (connectivity): an incidence from edges to meshes. The target modeling is similar to the radar: the receiver is depicted by 2 edges / impedances (Z_3 and Z_4) and the system impedance is named g_2. The whole system needs to gather radar and target ($g_1 \oplus g_2$) and include near and far field interactions. The impedances tensor is g, which gives the equations of the problem: $E = gk$, or tensorially writing: $E_\nu = g_{\nu\sigma}k^\sigma$ where k are the unknown mesh currents to solve and E stands for sources.

2.2. Tenfold

Assuming any system may be represented by a graph of networks, they are defined through their incidence and connectivity (giving the relations between N nodes, B edges and M meshes). The set (topology, T, impedance tensor, g, called the metrics and sources, E) gives all the information to describe a network or a system. These 3 previous objects are gathered in a list named "tenfold" (symbol $\breve{u}$, Fig.1(b)) associated with the set (T, g, E). Linking transformations to the tenfold enables transient topology evolutions; we may define transformers: $\breve{u}' = \breve{t} \cdot \breve{u}$ with $\breve{t} = (\Lambda,\ M,\ N)$. Mathematically, the evolution may be represented via gamma matrix [5].

3. Human factor

A system evolution can be entirely represented using a "tree model" (Fig.1 (b)), standing for the whole transformations. At this stage, one may wonder what is the reason to chose one particular transformation? Actually the result is the consequence of a decision depending of the so called Human Factor (HF). So it is natural to define the probability of making one choice compared to the others. Probabilities weights have to be affected to the different branches. This offers an introduction of HF. Game theory [6] details the interpretation of the evolution tree and may be convenient for communication networks [7, 8]. The information vector can be separate into two parts: first part still devoted to the state (tenfold), and second one is devoted to ways probability. It is automatically obtained through the gamma matrix also separated into two parts (transformations and probabilities of evolution). In this issue, thinking a solution is better to another may lead people to act differently. That is why the definition of a valuable payoff is crucial (the higher the gain is, the better the strategy of the gamer is). Once an output is computed at each step of evolution, the product of the probability to reach this state by the related gain provides useful information. The addition of all possible trajectories gives the average gain function of one player.

4. Numerical application

For demonstration purposes, the game mesh is sampled considering at maximum 10 steps (3 choices are proposed each time, see Fig.1 (a)).

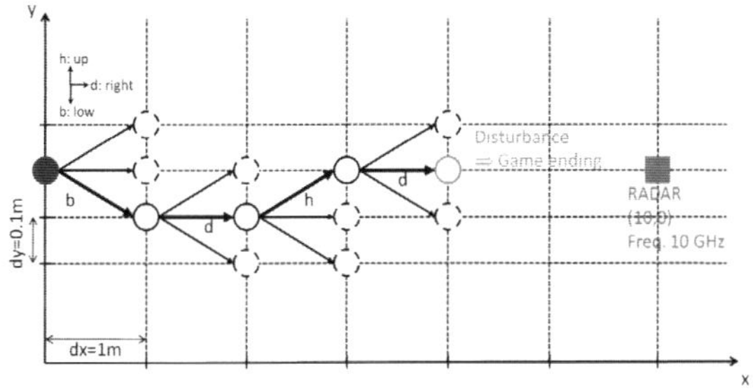

(a)

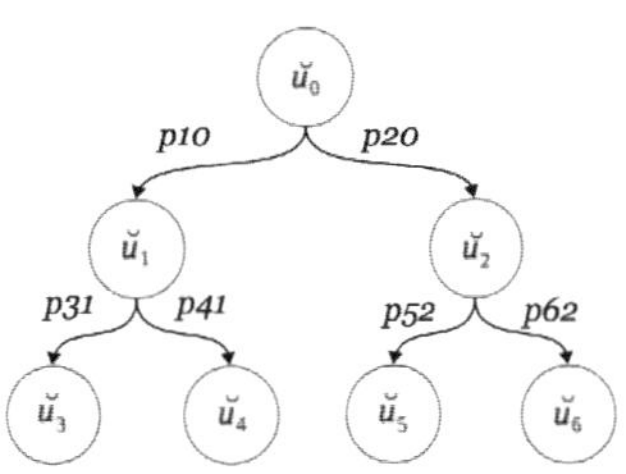

(b)

Figure 1. (a) Example of game path ("b d h d") and (b) Evolution tree with probabilities.

The equation of the problem is

$$GE = \frac{\left(\tilde{\gamma}_{n_e} \cdots \tilde{\gamma}_0 \tilde{I}_0\right)^{n_e}}{Distance\left(\gamma_{n_e} \cdots \gamma_0 \breve{u}_0\right)}, \tag{1}$$

where GE is the gain expectation, n_e the ending iteration of the game, and the $\tilde{\gamma}$ part of the propagator $\bar{\gamma}$ [4] is given from player's "psychological" profile.

4.1. Description

A radiation pattern is affected to each antenna, as antenna gains. The Green's function coupling electronics is

$$G(\theta,\ \xi) = G_1 G_2 cos\left(\theta\right)\left\{\frac{e^{-jkR}}{R} + \frac{e^{-jkR'}}{R'}cos\xi\right\}. \tag{2}$$

G_1 and G_2 stand for antenna gains. R and R' are respectively the distances, direct interaction and via ground reflection, separating moving target and RADAR. The angles θ and ξ are related to the target radiation pattern and to its angle with the ground; thus G_1 is fixed and G_2 depends on θ angle (given $G_2 = cos\left(\theta\right)^2 + 0.001$).

Three profiles are considered in this study: "go-getter" gamers, ones who are "undecided" and those whose path is entirely "random". For each kind of psychological profile a sequence is randomly or deterministically defined: it is a list of letters: "d" for right, "b" for down and "h" for up, see Fig.1 (a). Each time the player chooses one step beyond: he may go one step up, down or go on straightforward (right). After achieving a movement, one may compute the distance and angles between him and the radar and so compute the interaction between them. To this end, the tensorial equation detailed before ($E = gk$) is solved. Assuming the electronic device susceptibility, it is straightforward to compute its potential disturbance; depending on the result, the game continues or ends. The potential player's choices allow defining 2 different "classes": "straight" displacement ("bbb..." for instance) and "change of direction" (including "undecided" and "random" profiles). The "profiles" definitions rely on a priori 3 profiles: "go-getter" (3 paths straight along "d", "b" or "h" axis), "undecided" (3 different paths ups and downs through "b" and "h" axis) and purely "random". Each step of the game brings crucial information: the PG graph offers a current view of the system state via payoff computation. The probabilistic graphs obtained for each profile enables defining the best strategy to be as close as possible to the radar. The probabilities for the player to chose either "d", "b" or "h" rely on the given profile and are: "go-getter" $\rightarrow$ (0.8, 0.1, 0.1) (putting the focus on "d" trajectory), "undecided" $\rightarrow$ (0.25, 0.25, 0.5)(emphasing path oscillating around y = 0m) and "random"

$\rightarrow$ (1/3, 1/3, 1/3). The payoff gain is linked with the inverse of the distance to the radar when the game comes to an end.

4.2. Results and discussion

A profile always following the same group of direction (PG graph) is called "border trajectory" (case of "go-getter" gamers). "Undecided" and "random" trajectories are both characterized by "straight" and "change of direction" groups of transformations.

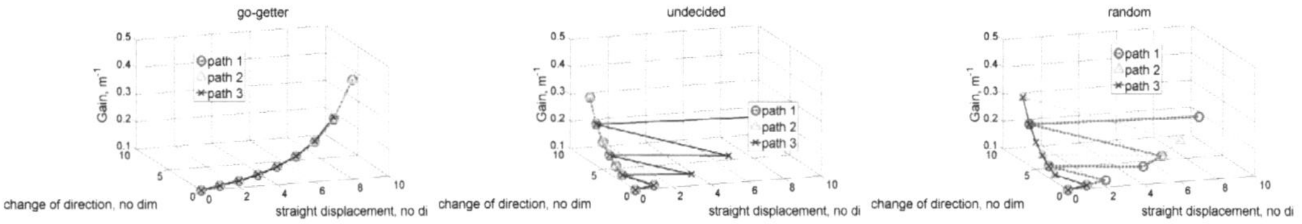

(a) (b) (c)

Figure 2. PG graphs from (a) "go-getter", (b) "undecided" and (c) "random" profiles.

First, from "go-getter" profile, the gamer follows a straight direction all the time and reach a highest payoff around 0.47. Figure 2 (a) shows the trajectories in the PG graph for the three possible sequences: "bbb...", "hhh..." and "ddd...". The expected gain is $GE = 0.07$ with a major part dedicated to third trajectory (payoff is 0.33) whereas first and second paths offer highest payoffs. Similarly to "go-getter" profile, the Fig.2 (b) depicts the paths from "undecided" case. It leads in the best case to a payoff of 0.3 with $GE = 0.00264$. Finally, three trajectories were randomly simulated and are depicted in Fig.2 (c). Since no a priori behavior is expected (random), GE is smaller than in previous cases ($GE = 0.0005$) but the payoff is very similar to "go-getter" and "undecided" strategies.

5. Conclusion

As depicted previously, gain expectations give information about the optimum decision. Among the defined strategy ("go-getter", "undecided", "random"), "go-getter" profile was the more interesting: the proposed methodology allowed to characterize the highest payoff with the smallest distance to the radar (2.15m). This cannot be intuitively obtained (radiation patterns influence and complexity of the Green's function could have favored "ups and downs" approach). Future works should improve the formalism (for instance partial payoff and reliability testing).

References

[1] Kondo K 1968 RAAG memoirs of the unifying study of basic problems in engineering and physical sciences by means of geometry (Vol. 4) *[Gakujutsu Bunken Fukyu-kai [Association for Science Documents Information]*
[2] Kron G 1942 A short course in tensor analysis *John Wiley and Sons* New York
[3] Maurice O, Reineix A, Hoffmann P, Pecqueux B and Pouliguen P 2011 A formalism to compute the electromagnetic compatibility of complex networks *Advances in Applied Science Research*, 2(5) 439–448
[4] Reineix A and Maurice O 2012 Modèle de gamma matrices pour la prise en compte d'interactions rayonnées entre réseaux *In Proc. 16th Int. Symp. on EMC* Rouen France
[5] Maurice O and Reineix A 2011 Proposition d'un formalisme comme support pour les études théoriques en systèmique *Systemica2011*
[6] Fudenberg D and Tirolle J 1991 Game Theory *MIT Press*
[7] Saad W, Han Z, Debbah M, Hjorungnes A and Basar T 2009 Coalitional game theory for communication networks *IEEE Signal Processing Magazine, Special Issue on Game Theory for Signal Processing and Communication*, 26(5) 77–97
[8] Rose L, Perlaza M, Lasaulce S and Debbah M 2011 Learning equilibria with partial information in decentralized wireless networks *IEEE Communication Magazine, Special Issue on Game Theory* 49(8)

Education and Public Outreach activities in Radio astronomy with the SKA South Africa.

N Oozeer[1,2,3], B A Bassett[2,4,5], K de Boer[1]

[1]SKA South Africa, The Park, Park Road, Pinelands, Cape Town 7405, South Africa.
[2]African Institute for Mathematical Sciences, 6-8 Melrose Road, Muizenberg 7945, South Africa
[3]Centre for Space Research, North-West University, Potchefstroom 2520, South Africa.
[4]Department of Mathematics and Applied Mathematics, University of Cape Town, Rondebosch, Cape Town, 7700, South Africa
[5]South African Astronomical Observatory, Observatory Road, Observatory, Cape Town, 7935, South Africa

E-mail: `nadeem@ska.ac.za`

Abstract. A Human Capital Development (HCD) program is a crucial part of any large organisation, and especially for large new research facilities such as the Square Kilometre Array (SKA) Africa. HCD provides a way of developing and channeling new minds into a very demanding field that ensures sustainability of the project and a multitude of spin-off benefits. Apart from educating learners at various levels, the HCD program must also inspire and educate the general public about the projects via an active outreach program. We highlight the various types of outreach activities that have been carried out in South Africa and the other SKA Africa partner countries. While there exist many teaching models we introduce and explore a novel concept of peer teaching for research known as the Joint Exchange Development Initiative (JEDI) and present some of its results. The JEDI workshops have resulted in a considerable number of learners embarking on advanced careers in science and research, and the demand is still growing.

1. Introduction

Outreach is a very important activity for an organisation especially for new upcoming facilities such as the Square Kilometre Array (SKA) Africa. Outreach is not only always about explaining a particular project but also at times it can be useful to channel more learners into science education, especially in Africa. There has been significant development in astronomy, in Africa (particularly in South Africa), in the past decades. The Southern African Large Telescope (SALT) and the High Energy Stereoscopic System (HESS) have been constructed and commissioned, in South Africa, and in Namibia respectively. South Africa is currently constructing the 64-antenna array radio telescope, MeerKAT, which will be incorporated into phase 1 of the SKA. In addition, nine other countries in Africa are participating in the SKA project, as well as the African Very Long Baseline Interferometer Network (AVN). These instruments prompt for a demand of skills and professional such as technicians, engineers and scientists have caused a huge quest for brain-gain and urge students to opt for science and engineering.

There are various ways of going through outreach in astronomy, science and other fields. For example it is very easy to put in a telescope and ask the learners to have a star gazing party. It is also very attractive to show colourful optical celestial images, but showing radio contours and false colour pictures of radio sources is not so attractive. It is not very straightforward when has to do outreach in radio astronomy and talking to learners and the general public about it. Lack of visual memory can affect remembering and consequently learning in many people. The fact that nowadays majority of the learners know and have access to huge amount of information, this can help us to do radio astronomy outreach effectively.

This paper is divided as follows: §2 discusses the various types of outreach activities that have been carried out in South Africa and §3 describes the Joint Exchange Development Initiative (JEDI), that is one of the education human capital development (HCD) programme that we have implemented. We finally conclude the paper in §4.

2. Outreach overview

Before going into detail about radio astronomy outreach, we shall try to review some of the key points generally taken into account during outreach events. A question we always ask is what do we want to convey during outreach. We are already submerged by so much of information nowadays so why will people still want to hear about us. Many will agree that knowing the audience is a key point for efficient outreach. We can classify the target audience for outreach as easily reachable and hard-to-reach. Sandu et al. (2011) [1] mention, how can we grab the attention of all those targets that have not yet been converted, especially the hard-to-reach.

Timing is very crucial while going for outreach. It might not be wise to do outreach to students during the pre-examination period for they will never want to listen. On the other hand, catching students when they will be making their subject choices can be very effective. An important role in outreach is to trigger the wow factor and get the learners interested and also leaving them with this awe that will make them quest for more.

2.1. Onsite visit

Onsite visits consist of going to various schools, community centres, and science festivals to talk about our project. During these visits, talks and presentation are the best way to reach the audience. However, this also involves a mixed crowd and one shall be careful as it can have a negative impact if one is not well prepared and can kill the essence of outreach. It is advisable to rework ones outreach methodology according to the crowd. Animated presentations with little or no scientific equations does help a lot for general public. Posters/banners are very useful to attract attention if they are very colourful and nicely presented. Freebies always attract lots of people to a stand for example, but the quality and quantity can leave many people unhappy.

2.2. Mass media

Mass communication is another aspect of outreach. We can reach a huge number of people in a very short time span. However, mass media has evolved drastically since the last decades. We are moving towards a computerised electronic communication that has its advantages as long as the target group has access to such technology. Talks on the radio tend to attract elderly groups while career related television shows have more impact on the younger generation.

2.2.1. Audio/Video Audiovisual has always been the privileged way to communicate to the mass. However, it is quite difficult to ensure that all the audience has got the right information. One key point is to make small series of a programme, each time explaining a small portion of a project. Nowadays with social media and the Internet, lot of outreach activities can be done using these media. There exist many other outreach activities similar to the one we are doing. Using these outreach materials into ones (acknowledging the various authors) event allows strengthening the outreach materials and also show the diversity of similar projects. A project web or wiki page written in diverse languages allow a good flow of information from the source to the mass. However, designing of a good web page is also very essential in order not to make the Internet users lost in a sea of information.

2.2.2. Printed media Printed media such as; newsletter, newspaper, flyers are still very useful to reach a different crowd. Even though we have moved into an electronic version of all these form of media, in developing countries there are lots of places where there still exist no Internet facilities or audio/visual equipment. In these places, printed materials coupled with hands on activities have shown to be very useful. The SKA Africa came up with the Mission MeerKAT cartoon to explain the SKA project. This project has reached a huge crowd and when one revisits a place for outreach, very often the learners who heard of these cartoons would want their own copy while others are eager to get the next volume. Starting with the concept of explaining what is the SKA Africa project, this has evolved into a more pedagogical tool to teach students about other SKA partner countries, from the extinct dodo of Mauritius to national animals of these countries and the various instruments to be used in the partner countries.

2.3. Artistic approach

Artistic shows and performances can be used to bridge the gap between culture and science. Archeo-astronomy, the study of how various civilisations perceive the sky and the Universe is now more frequently discussed and form an interesting field of research. From the Australian aborigine view of how the Milky Way [2], was created by the emu to the seven sisters/princesses in African folk stories [3] to talk about the Pleiades, astronomy has been well been used by the ancients. It is an innovative way to awaken people imagination about various scientific issues. However one has to be very careful not to touch the sensitivity of one belief and cultures.

2.4. Radio astronomy outreach

It can sound very complicated when one start talking about signals, filters, amplifier and other technical terms. One starting point for radio astronomy is to use the "Eye and I" concept. The human eye collects visible light and the lens brings the light to the retina where it will be digitised and transferred to the brain. The "I" concept questions the target audience about what they are already know and use in their everyday life and then bring it to another level.

In brief, a radio telescope consists of various parts. The dish that captures the signal coming from outer space, reflects it to a sub-reflector, converts the signal into electrical signals that is then transmitted to the computing system where the signal is further analysed. A digital satellite dish can be used to teach radio astronomy at various levels. The output from the receiver of a satellite dish can be connected to a satellite finder, which has an output to a speaker. The output can also be connected to the microphone input of a computer running specific software and can now be recorded, listened to and displayed. The students can see the radio signal. Similarly the same dish can be used to measure signal from various sources and detect radio frequency interferences (RFIs).

3. Education with JEDI

In education, the curriculum provides a plan sets for learning, while the instruction is the actual engagement of the learners with planned learning opportunities. The teaching is a process whereby one mediates between another person and the materials. There exists many teaching models from lectures, discussion, questioning, to independent learning and self-teaching. The Joint Exchange Development Initiative (JEDI - http://jedi.saao.ac.za) is a way to maximise on the teaching methodologies.

The JEDI is a concept to enhance development and education via direct transfer of skills and expertise in any specific field. It is an initiative to provide development via joint exchange among stakeholders. This is achieved by bringing stakeholders: students, post-docs and staffs together in an informal but intense research environment to tackle unsolved problems for e.g. in Astronomy. [BB] introduced the concept in South Africa and since then various models/variants have been tried and tested depending on the need and requirements of the participants. From these interactions, scientific publications [4, 5, 6, 7, 8] have been produced and stronger collaborations have been created. The JEDI concept is based on the following insights about research and learning:

- It is well known that true learning occurs when students actively engage with material. Listening is not active engagement. Hence most conference talks do not provide real learning environments. Research success requires skills in the most relevant topics and toolsets. In the 3rd world, supervisors are often out of date in terms of the topics they research and the tools they use.
- Conferences advertise the result of years of research; they do not teach participants how to do research. It is the lack of examples and role models of how to do good research that is most critically absent and most desperately needed in the third world.
- Ability to work in teams is critical to modern research success. Yet many undergraduate and graduate programs, especially in third world countries, are mired in education approaches based on individual work. Most students lack the interpersonal skills to work and communicate effectively in teams.

The JEDI model address all three of these major issues at one stroke. It is based on active engagement of real research projects with other students and faculty. It fosters friendships, teamwork and community between the students and with the faculty which all persist years after a workshop. The six JEDI workshops that have been held have shown just how successfully this is achieved. These successful models are now replicated to other African countries (Mauritius, Kenya, Namibia, etc.), especially in the future era of huge datasets with the SKA.

4. Conclusion

In this paper we have reviewed some of the outreach activities that we have initiated and continuing with the SKA Africa. It is a way of channeling new brains into a will be demanding area. Such feeder will eventually sustain the project. We also presented the novel method of peer teaching through the JEDI. These activities are producing their fruits with more and more students are moving into science and engineering. These activities are undoubtedly producing new generation Astronomers and Engineers for the upcoming SKA Africa.

5. Acknowledgments

NO would like to acknowledge assistance, support and encouragement from the SKA Africa colleagues who have contributed enormously in various activities related to this work.

References

[1] Sandu O. and Christensen L. L., Outrageous outreach, CAP Journal, July, 2011.
[2] Norris, R. P., & Hamacher, D. W. 2011, IAU Symposium, 260, 39
[3] Lloyd Bleek Collection, University of Cape Town (2007), http://lloydbleekcollection.cs.uct.ac.za
[4] Bassett, B. A., Brownstone, M., Cardoso, A., et al. 2008, JCAP, 7, 7
[5] Crawford, S. M., Ratsimbazafy, A. L., Cress, C. M., et al. 2010, MNRAS, 406, 2569
[6] Newling, J., Varughese, M., Bassett, B., et al. 2011, MNRAS, 414, 1987
[7] Kessler, R., Bassett, B., Belov, P., et al. 2010, PASP, 122, 1415
[8] Zwart, J. T. L., Jarvis, M. J., Deane, R. P., et al. 2014, MNRAS, 439, 1459

IOP Conf. Series: Materials Science and Engineering **67** (2014) 012017 doi:10.1088/1757-899X/67/1/012017

Big Data analytics and cognitive computing – future opportunities for astronomical research

M.A. Garrett[1,2]
[1]ASTRON, the Netherlands Institute for Radio Astronomy, Postbox 2, 7990 AA Dwingeloo, The Netherlands
[2]University of Leiden, Leiden Observatory, Postbus 9513, 2300 RA Leiden, The Netherlands.

E-mail: garrett@astron.nl

Abstract. The days of the lone astronomer with his optical telescope and photographic plates are long gone: Astronomy in 2025 will not only be multi-wavelength, but multi-messenger, and dominated by huge data sets and matching data rates. Catalogues listing detailed properties of billions of objects will in themselves require a new industrial-scale approach to scientific discovery, requiring the latest techniques of advanced data analytics and an early engagement with the first generation of cognitive computing systems. Astronomers have the opportunity to be early adopters of these new technologies and methodologies – the impact can be profound and highly beneficial to effecting rapid progress in the field. Areas such as SETI research might favourably benefit from cognitive intelligence that does not rely on human bias and preconceptions.

1. Introduction

Astronomical facilities, such as the new generation of radio telescopes e.g. LOFAR [1] are often considered as good examples of the type of Big Data generators that are rapidly emerging in a wide variety of different scientific fields. Telescope facilities like LOFAR can routinely produce data sets of ~ 100 TB per day, and new facilities such as the SKA [2] will generate object catalogues that contain billions of sources [3]. The limited ability to process these huge data sets and make sense of the complexity of the archived data products is already emerging as a major bottleneck in the scientific exploitation of these new instruments.

Initiatives such as DOME, the ASTRON IBM Center for Exascale Technology [4] are addressing some of these technical issues by investing significant effort in areas that include low-power (green) data streaming platforms, exascale storage solutions and the extensive implementation of nano-photonics. The challenge of the SKA forces much of the work to concentrate on realising systems that can deliver much higher performance, on a reduced footprint with lower power consumption. These efforts mostly focus on front-end hardware and back-end data challenges but there is a clear ambition to begin to extend this work into the post-processing domain via advanced Big Data analytics. In particular, the next generation of all-sky astronomical surveys are set to generate a total data archive that also approaches Petabyte scales. It is no longer possible to consider that a single astronomer, or a large scientific team, or even a global citizen scientist project can fully exploit the scientific potential of these data sets.

Astronomers are in a good position to benefit from the nature of their data and the extreme scale of their problem - telescopes like LOFAR and SKA make for interesting test beds for the development of novel techniques and advanced systems that can have additional value in many other fields, including those with significant societal benefit via commercial and industrial application. By helping to develop and become early adopters of these advances, astronomers have the opportunity to make great strides in fully exploiting advances in data analytics and cognitive computing in order to fully exploit the huge catalogues of multi-dimensional, multi-wavelength and multi-messenger data coming their way. The DOME collaboration is particularly well placed to further develop these ideas - in this paper we consider how these efforts might develop over the next 10-15 years.

2. Big Data and advanced analytics for astronomy

Big Data is usually defined in terms of the three V's – Volume, Velocity and Variety [5]. All three characteristics apply to astronomical data – the first two are obvious (large data sets and data streaming or data "in motion"), and the third "Variety" relates to the diversity of data type – in astronomy this can be the considered to relate to the myriad of different instruments, wavelength range/messenger type recorded, and data format. A fourth V, "Veracity" is important in all areas of analytics, and is well known to astronomers dealing with measurements that are inherently uncertain (due to limits of the instrumentation) or possibly corrupt (e.g. due to the presence of man-made interference in radio astronomy data sets).

While it is true that telescopes like LOFAR are prodigious generators of the 4 V's, the amount of data being generated around the planet is also quite staggering. It's estimated for example, that Google currently processes ~ 25 PB of data every day, and that ~ 3 EB of data are generated every day around the world. The Internet itself is an impressive beast with a total storage space of ~ 250 EB, total data rates of ~ 8TB/second and a total cost estimated at 5% of the world's total power requirements. Cloud storage systems already count for 1 EB of the total storage capacity, and this capacity is set to grow while the costs reduce. By the end of the decade, this data avalanche enhanced by the growth of sensor networks (of which LOFAR is one of the first multi-disciplinary examples) is expected to increase 30-fold.

Analysing the combination of traditional data sources with unstructured data (e.g. social media) is today's big business, and the investments being made by large commercial companies are significant. Many of the advances now being made via a variety of Big Data initiatives are likely to be applicable to astronomy – not only the distributed server hardware but also the underlying software infrastructure e.g. scalable open-source software such as Hadoop, supporting distributed processing of large data sets across clusters of commodity servers. In order to understand and predict the behavior of customers, forecast financial trends, or uncover operational insights that can increase efficiency, huge investments are being made in areas such as data analytics. Often, correlations and patterns in the data must be identified quickly – usually in real-time, if they are to realise their full commercial impact. Brute force correlations of all data types are leading to new and surprising insights that were not intuitively obvious from the start. Naturally, it's important not to confuse correlation with causality but a brute force approach, coupled with sophisticated pattern recognition algorithms can reveal hard-to-spot connections.

As the size of astronomical survey data sets is set to grow considerably over the next decade, an approach based on commercial data analytics may offer considerable advantages. For example, SKA sky surveys will include billions of sources, and each source will be described and characterized by many different parameters including location, size, morphology, flux density, spectral index, line

IOP Conf. Series: Materials Science and Engineering **67** (2014) 012017 doi:10.1088/1757-899X/67/1/012017

width, redshift etc. Combining this information with comparable surveys at many other wavelengths (a veritable astronomy mashup), requires a new approach to large-scale data analysis, and there can be little doubt that the techniques now being pursued commercially in the realm of Big Data can also be applicable to the astronomy case. In addition to advanced pattern recognition, algorithms working on streaming data can be specifically designed to identify anomalies or outliers, and to preserve them (rather than the case now where they are often thrown away). This can be important in enabling serendipitous discovery relevant to a wide-range of studies but perhaps Transients and SETI (Search for Extraterrestrial Intelligence) research stand to benefit most.

Another area where commercial investment in data analytics is making great strides is in data visualization. Progress in data visualisation has been at a virtual standstill for the last few decades in astronomy, for example, spectral line cube data sets remain cumbersome to inspect and explore. With data analytics we can hope to use new visualization approaches, featuring automatic feature extraction and collaborative sharing tools. Finally, predictive analytics (and in principle also prescriptive analytics) must also have something to teach astronomers, especially in the area of regular variables or repeaters, and again transients.

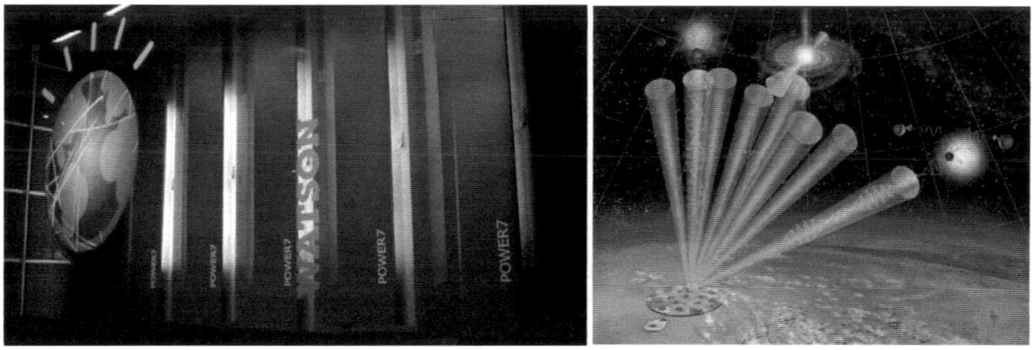

Figure 1: Applying cognitive computing to large astronomical surveys could have a huge impact in terms of serendipitous discovery - areas that might specifically benefit include those like SETI where human bias and other pre-conceptions may limit current efforts.

3. Cognitive computing for astronomy

Cognitive computing seeks to leap beyond the deterministic boundaries of conventional programmable (von Neumann) computers by introducing dynamic systems that learn from the data, sense their environment, place unstructured data such as video in context, independently pursue new courses of action and directly interact with humans by understanding and using natural language. IBM's Watson machine [6] represents an important step forwards towards this goal but further progress requires revolutionary new architectures to be developed that mimic the human brain in which memory and processing power are tightly integrated into sub-units that are distributed, run in parallel and communicate (both internally and externally) via nano-photonic connectors [7].

The impact of such systems, guided by experienced scientists would surely be profound. At least initially, the astronomer's human qualities of intuition, insight and creativity would be important guiding inputs to the direction this new discovery process might take. Many fields of science stand to benefit, especially those that are data dominated. Naturally there are also risks involved in developing technologies that approach or perhaps even exceed human capacity. But this is the slippery path on which we are all travelling – natural and good communication between machines and humans will be vital element towards a successful outcome.

4. Conclusions

Over the last decade, radio astronomy has successfully embraced COTS (Commercial off-the-shelf) technology with the new generation of telescopes like LOFAR making good use of the many advances fuelled by the global investment in ICT. The huge volume and increased complexity of future large sky surveys, coupled with the emergence of multi-messenger facilities, demand a new approach to scientific discovery in astronomy (but also other fields). An opportunity exists to make great leaps forward by engaging with the Big Data programme, re-using the rapidly developing suite of data analytics (and the associated visualisation techniques). This approach can greatly benefit the full scientific exploitation of large astronomical surveys, leading to new discoveries on accelerated timescales.

The great promise of cognitive computing is also on the horizon – this could propel astronomical research in very new and perhaps unexpected directions. Projects like DOME are important vehicles for ensuring early engagement of the astronomy community with cognitive systems. In particular, areas such as SETI research might really benefit from an approach that is largely independent of human preconceptions, potentially leading to game-changing discoveries for all mankind.

References

[1] van Haarlem M P, Wise M W, Gunst A W et al. 2013, *Astronomy & Astrophysics* **556** A2
[2] Dewdney P 2013 SKA1 System Baseline Design *SKA Project Documents*
[3] Wise M W, Alexov A, Folk M et al. 2011 Towards HDF5: Encapsulation of Large and/or Complex Astronomical Data, *Astronomical Data Analysis Software and Systems XX. ASP Conference Proceedings vol 442* ed Ian N. Evans, Alberto Accomazzi, Douglas J. Mink, and Arnold H. Rots (San Francisco: Astronomical Society of the Pacific) p 663
[4] Broekema P C, van Nieuwpoort R V & Bal, H E 2012 *Proceedings of the 2012 workshop on High-Performance Computing for Astronomy* ACM pp 9-16
[5] Laney D 2001 3D Data Management: Controlling Data Volume, Velocity, and Variety *META Group technical report*
[6] Ferrucci D, Levas A, Bagchi S, Gondek D, Mueller E T 2013 *Artificial Intelligence 199-200* (Amsterdam Elsevier) pp 93-105
[7] Kelly J E and Hamm S 2013 *Smart Machines* (New York: Columbia University Press)

Acknowledgments

I would like to acknowledge the DOME project. Attendance of the author at Radio 2014 was partially supported via an IBM Faculty Award.

IOP Conf. Series: Materials Science and Engineering **67** (2014) 012018 doi:10.1088/1757-899X/67/1/012018

The Science and Technology of the Square Kilometre Array

S.J. Tingay[1]
International Centre for Radio Astronomy Research, Curtin University
Bentley, Western Australia, Australia, 6102

E-mail: s.tingay@curtin.edu.au

Abstract. I discuss the science and engineering of the Square Kilometre Array (SKA), the next generation astronomical observatory for studying the structure and evolution of the Universe. I give an overview of the SKA Key Science areas and map them onto the instruments that are being developed for these investigations. I outline where the SKA is being built and how the international community is developing the project. I discuss a range of SKA pathfinder and precursor telescopes that are currently being built or are operational, helping to inform SKA engineering, ICT and astrophysics.

1. Introduction

This article follows the keynote presentation I gave at the "RADIO2014" conference. The presentation was a broad overview of Square Kilometre Array (SKA) science and engineering and, likewise, this written account is broad and shallow, given the limited length of the article.

When we humans gaze at the night sky, our eyes detect photons in a very narrow wavelength range, approximately 650 to 1200 nanometres. Consider the image in Figure 1, a photograph of the southern Milky Way (our Galaxy), which reveals huge amounts of gas, dust and stars in our Galaxy due to thermal emission processes associated with these objects.

As spectacular as Figure 1 is, it only reveals one small part of the information available to astronomers. Other parts of the electromagnetic spectrum, such as in the high energy photon regime (X-ray and gamma-ray) and low energy photon regime (radio waves) give very different information for the same objects. For example, when we look at the southern Milky Way in X-rays we see high-energy particles and processes such as the accretion of material onto Black Holes and other compact objects. At the other end of the electromagnetic spectrum, many objects and processes produce radio waves (wavelengths greater than approximately 1 millimetre) that can be studied using radio telescopes.

Modern radio telescopes are configured as arrays of individual elements (either small dish elements or low frequency antennas) and used as interferometers. See [1] for an explanation of interferometry. In this way, radio telescopes can be built with large collecting areas (and thus high sensitivity) with elements spread over large geographic regions (leading to high angular resolution imaging). Instruments such as the Australia Telescope Compact Array (ATCA: [2]) and the Low Frequency ARay (LOFAR: [3]) are examples of multi-element interferometers using dishes and low frequency dipoles, respectively.

[1] To whom any correspondence should be addressed.

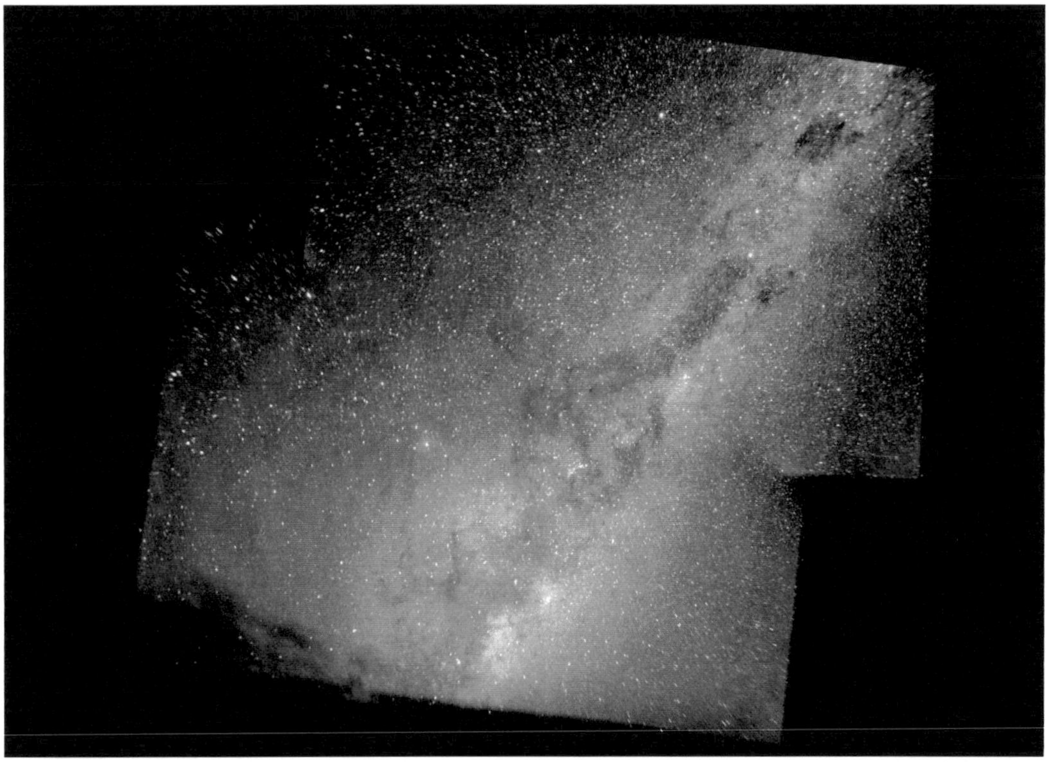

Figure 1: The southern Milky Way, as seen from Australia's SKA site in Western Australia, at wavelengths of light to which human eyes are sensitive (credit: the author).

The Square Kilometre Array (SKA) takes the idea of interferometry into new territory, with the construction of radio telescopes consisting of hundreds to millions of individual elements spread over hundreds to thousands of square kilometres.

2. The Square Kilometre Array (SKA) – the science

The SKA is a project to construct the largest and most sensitive radio telescope ever built, supported by an international community of 10 countries. A great deal of information on the SKA project, including material covered in summary form below, is available on the SKA web pages: http://www.skatelescope.org.

The science of the SKA is aimed at an exploration of fundamental astrophysics and physics and falls into five main areas:

(1) How do galaxies evolve? What is Dark Energy? Only approximately 5% of the content of the Universe is made up of baryonic material. The remaining 95% of the Universe is made up of Dark Matter (~25%) and Dark Energy (~70%), for which physics has no generally accepted explanation. The SKA will attempt to investigate the properties of Dark Matter and Dark Energy by watching how galaxies evolve over cosmic time, tracing the effects of Dark Matter in galaxies and by tracing the geometry of space-time and its relationship to Dark Energy.

(2) Was Einstein right about gravity? The SKA will be used to hunt for new pulsars, the rapidly spinning remnants of massive stars at the end of their lives. Pulsars have extremely strong gravitational fields and can be used as laboratories for tests of theories of gravity, including the highly successful general theory of relativity.

(3) What generates giant magnetic fields in space? Magnetic fields exist on all spatial scales in nature, including on the largest scales of hundreds of thousands or millions of light years (the scales of galaxies or clusters of galaxies). The origin of these magnetic fields is unclear as is their relationship to magnetic fields on much smaller scales. The SKA will be able to examine magnetic fields on all astrophysical scales, including the largest scales.

(4) How were the first stars and black holes formed? After the Big Bang, the Universe cooled to the point where neutral hydrogen filled the Universe. Over the first billion years of the Universe, this gas, acting under the influence of gravity, likely caused by Dark Matter, formed into the first luminous objects – stars, black holes and galaxies. Using low frequency radio waves, the SKA can look back in time into the first billion years to provide a unique view of this era in the Universe's history.

(5) Are we alone? The SKA will be sensitive enough to detect stray radio emission from communications systems that may exist on other planets, assuming that those transmitters are similar to those developed by humans on earth. Thus, the SKA will provide the best yet search for extra-terrestrial signals. Further, the SKA will be used to explore the process of planetary system formation and studies of astrobiology, some of the fundamental building blocks of life (as we know it) in the Universe.

All of these science drivers require very high sensitivity and very good angular resolution over a substantial range in radio frequency, from ~100 MHz to ~10 GHz. As such, the science requirements translate into a very challenging set of technical requirements. The next section of this paper briefly discusses the technology required for the SKA.

3. The Square Kilometre Array (SKA) – the technology

The SKA will be built in two locations, Southern Africa and Western Australia. Both locations have been chosen because they have very low levels of interference from human activities that produce radio waves (mobile phones, telecommunications, industry activity, automobiles etc) many orders of magnitude stronger than the astronomical signals sought. Both these locations are therefore remote.

The SKA will also be built in two stages, Phase 1 and Phase 2. Phase 1 is nominally approximately 10% of the full SKA scope, with the remaining 90% slated for construction in Phase 2. The breakdown of SKA construction is as follows:

Southern Africa, Phase 1: An array of approximately 250 dishes of order 13 m diameter, operating in the 0.35 – 3.0 GHz frequency range, will be built using traditional single pixel feed technologies. This telescope is known as SKA-mid.

Western Australia, Phase 1: An array of approximately 100 dishes of order 12 m diameter, operating in the 0.65 – 1.7 GHz frequency range, will be built using a new technology known as Phased Array Feeds (PAFs). Whereas single pixel feeds give a single field of view, PAFs provide a large number of independent fields of view, covering a large area of sky compared to a comparable single pixel feed telescope. This telescope is known as SKA-survey.

Additionally in Western Australia, a low frequency radio telescope will be built using approximately 200,000 dipole antennas operating in the frequency range 0.05 – 0.35 GHz. The low frequency telescope is inherently a very large field of view instrument and is known as SKA-low.

Southern Africa, Phase 2: SKA-mid will be expended to approximately 2,700 dishes. An additional instrument will be built in Phase 2, consisting of approximately 250 stations made of dense aperture arrays, similar to the low frequency array but at higher frequencies.

Western Australia, Phase 2: SKA-low will be expanded to approximately 2,000,000 dipole antennas.

The exact specifications for each telescope are to be determined over the next few years in the final SKA design stage. Each of these instruments are of a scale never before constructed and offer technical challenges across electrical and electronic engineering (RF and digital systems), power engineering (provision of electricity to significant remote infrastructure), civil engineering, information and communication technologies (computing and advanced networking), and algorithm development.

The scale of the SKA is such that we do not embark directly upon its construction. Three official SKA Precursors are being built on the two SKA sites. In Southern Africa, MeerKAT is being built as a Precursor for SKA-mid. In Western Australia, the Australian SKA Precursor (ASKAP: [4]) is being built as a Precursor for SKA-survey and the Murchison Widefield Array (MWA: [5]) has been built as a Precursor for SKA-low and is currently the only one of the three Precursors to be operational for science. These Precursors are proving the science and technology for the SKA, in advance of SKA construction. An additional SKA Pathfinder not located at one of the SKA sites is LOFAR, based in The Netherlands.

Finally, in preparation for SKA construction, the international community is funded to undertake the final coordinated SKA design and prototyping activities, an official project phase known as pre-construction. Pre-construction will continue for three or four years, resulting in tender-ready specifications for SKA construction. Pre-construction activities make heavy use of industry collaborators and the eventual SKA construction will engage industry at a fundamental level, as the scale of construction is well beyond the capabilities of astronomical research institutes.

4. Conclusions

The SKA is the biggest ever project in radio astronomy and one of the largest mega-science projects currently being pursued around the world; it is an international endeavour to pursue fundamental physics and will have many technological spin-offs that are likely to have significant impact outside astronomy or physics.

5. References

[1] Thompson, A.R., Moran, J.M. & Swenson Jr, G.W., 2001, Interferometry and Synthesis in Radio Astronomy, 2nd Edition (Wiley). ISBN: 978-0-471-25492-8

[2] Journal of Electrical and Electronics Engineering, Australia, Special Issue, 1992, Vol. 12, No. 2

[3] van Haarlem, M.P. et al. 2013, LOFAR: The LOw-Frequency Array, Astronomy and Astrophysics, 556, 2

[4] Johnston, S. et al. 2008, Science with ASKAP. The Australian Square-Kilometre Array Pathfinder, Experimental Astronomy, Vol. 22, Issue 3, 151

[5] Tingay, S.J. et al. 2013, The Murchison Widefield Array: The Square Kilometre Array Precursor at Low Radio Frequencies, Publications of the Astronomical Society of Australia, 30, 7

Acknowledgments

This scientific work makes use of the Murchison Radio-astronomy Observatory, operated by CSIRO. We acknowledge the Wajarri Yamatji people as the traditional owners of the Observatory site. This work has been supported by the International Centre for Radio Astronomy Research (ICRAR), a Joint Venture of Curtin University and The University of Western Australia, funded by the Western Australian State government.

Uncertainty evaluation in numerical modeling of complex devices

X. Cheng and V. Monebhurrun

Department of Electromagnetics, DRE-L2S, SUPELEC, 3 rue Joliot-Curie, 91192 Gif-sur-Yvette Cedex, France

E-mail: xi.cheng@lss.supelec.fr

Abstract. Numerical simulation is an efficient tool for exploring and understanding the physics of complex devices, e.g. mobile phones. For meaningful results, it is important to evaluate the uncertainty of the numerical simulation. Uncertainty quantification in specific absorption rate (SAR) calculation using a full computer-aided design (CAD) mobile phone model is a challenging task. Since a typical SAR numerical simulation is computationally expensive, the traditional Monte Carlo (MC) simulation method proves inadequate. The unscented transformation (UT) is an alternative and numerically efficient method herein investigated to evaluate the uncertainty in the SAR calculation using the realistic models of two commercially available mobile phones. The electromagnetic simulation process is modeled as a nonlinear mapping with the uncertainty in the inputs e.g. the relative permittivity values of the mobile phone material properties, inducing an uncertainty in the output, e.g. the peak spatial-average SAR value.The numerical simulation results demonstrate that UT may be a potential candidate for the uncertainty quantification in SAR calculations since only a few simulations are necessary to obtain results similar to those obtained after hundreds or thousands of MC simulations.

1. Introduction

Whenever a measurement is performed, it is usual practice to evaluate the uncertainty for which international guidelines have been previously elaborated [1]. For example, when a measurement instrument is calibrated, the uncertainty is formally stated in the measurement report. When a numerical result is derived using an electromagnetic simulation tool, the associated uncertainty is rarely indicated. Uncertainty is obviously associated with any numerical simulation process e.g. model or structural uncertainty, parameter uncertainty etc. Uncertainty quantification in computational fluid dynamics has already been investigated during the past decades [2, 3].

Herein the focus is on the uncertainty quantification in numerical dosimetry or specific absorption rate (SAR) calculations using mobile phones and head models. Currently the SAR compliance of mobile phones is based on measurement procedures described in international standards [4]. Since the SAR compliance test of a mobile phone is costly and time-consuming, the availability of intuitive electromagnetic simulation software running on fast and affordable computers offers interesting perspectives for the SAR calculation of mobile phones, useful for pre-compliance tests, for example. A standardized procedure for the SAR calculationof mobile phones using the finite difference time domain (FDTD) method is currently being developed by a working group of ICES/IEEE/TC34 [5, 6]. A benchmark computer-aided design (CAD) mobile phone model was recently proposed for an international interlaboratory comparison of electromagnetic simulation tools [7]. Although there was a good overall agreement among the nine laboratory results, a maximum deviation of about 20% was observed in the peak 10g average

SAR value. Since the same phone model and the same material parameters were used by the different participating laboratories, sometimes using the same software, the differences in the applied mesh densities and positioning against the head phantom may explain the observed deviations. For comparable results, it is therefore crucial to state the uncertainty of the SAR calculation.

Uncertainty quantification in numerical dosimetry is not straightforward. For example, the traditional Monte Carlo (MC) [9] simulation method cannot be applied because a typical SAR calculation of a mobile phone placed against the specific anthropomorphic mannequin (SAM) head phantom requires about half an hour using a graphical processing unit (GPU) to several hours using a standard workstation. The unscented transformation (UT) [8,10-11] is an alternative and numerically efficient method herein investigated to evaluate the uncertainty in the SAR calculation. The electromagnetic simulation process is then modeled as a nonlinear mapping with the uncertainty in the inputs e.g. the relative permittivity values of the mobile phone material properties, inducing an uncertainty in the output, e.g. the peak spatial-average SAR value. Since UT may be applied non-intrusively, the electromagnetic simulation tool is therefore considered as a black-box.

2. Unscented Transformation (UT)

UT is basically the application of a nonlinear transformation to a set of discrete points to obtain the statistics of an unknown probability distribution [8]. Considering a nonlinear mapping f of a random variable x with mean $\overline{x}$ and variance σ_x^2, the output y of the nonlinear transformation is simply expressed as:

$$y = f(x) \tag{1}$$

and the objective is to obtain the statistics of y, such as the mean $\overline{y}$ and variance σ_y^2, from which the standard deviation (or uncertainty with a given confidence interval) is derived. The nonlinear transformation f is applied to a set of points, so-called sigma points x_i, yielding a corresponding set of transformed points y_i that are employed to obtain the statistics of y. As follows:

$$\overline{y} = \sum_{i=0}^{n} w_i f(x_i) = \sum_{i=0}^{n} w_i y_i \tag{2}$$

$$\sigma_y^2 = \sum_{i=0}^{n} w_i (f(x_i) - \overline{y})^2 = \sum_{i=0}^{n} w_i (y_i - \overline{y})^2 \tag{3}$$

where w_i are the weights which obey the condition $\sum_{i=0}^{n} w_i = 1$ and n is the order of UT. In essence, UT is rather similar to MC. Since both are used to find an approximation of the probability distribution function rather than the nonlinear function f which is thus seen as a black box. Nonetheless, there is a fundamental difference between these two non-intrusive methods: in the case of MC, a random sampling–in fact, an optimum sampling method such as the Latin Hypercube Sampling (LHS)–is applied to obtain the set of points of the variable x, whereas in the case of UT, the sigma points are deterministic. They are derived from the statistics or moments of the random variable and there are three main methods to calculate these sigma points from [10]: (a) symmetric method, (b) reduced method, and (c) spherical method. In [11], a modified version of UT based on the symmetric method was applied in electromagnetic simulation using the transmission line matrix (TLM) method. The same modified symmetric method is herein applied in conjunction with the finite integral technique (FIT) method [12]. The above two equations are then expressed as follows:

$$\overline{y} = \sum_{i=0}^{n} w_i f(\overline{x} + S_i) \tag{4}$$

$$\sigma_y^2 = \sum_{i=0}^{n} w_i (f(\overline{x} + S_i) - \overline{y})^2 \tag{5}$$

where S_i is the sigma point and $x = \overline{x} + \hat{x}$. $\hat{x}$ is a random variable with zero mean and known probability distribution. S_i are chosen from the distribution of $\hat{x}$ in the modified UT. The sigma points and weights are calculated as follows:

$$\sum_i w_i S_i^p = E[\hat{x}^p] \tag{6}$$

$$\sum_i w_i = 1 \tag{7}$$

where p is the order of approximation.

The above formulas can be extended to the case of two random variables, as follows:

$$\overline{y} = \sum_{i=0}^{n}\sum_{j=0}^{m} w_i w_j f(\overline{x_1} + S_i, \overline{x_2} + S_j) \tag{8}$$

$$\sigma_y^2 = \sum_{i=0}^{n}\sum_{j=0}^{m} w_i w_j (f(\overline{x_1} + S_i, \overline{x_2} + S_j) - \overline{y})^2 \tag{9}$$

where x_1 and x_2 are two random variables with mean $\overline{x_1}$, $\overline{x_2}$ respectively. S_i and S_j are sigma points of the two inputs.

3. Numerical modeling

A commercial package of FIT is herein used for the numerical simulations [12]. Two different commercially available dual-band mobile phones -900 MHz and 1800 MHz frequency bands- are investigated (Figure 1(a) and (b)). Both mobile phones were previously validated both numerically and experimentally. The mobile phone shown in Figure 1(b) was previously proposed as benchmark for the international interlaboratory comparison [7]. Since the SAR calculation with the SAM phantom (Figure 1(c)) is computationally expensive using the MC method, the efficiency of UT is first demonstrated by comparing with the MC results obtained for the case of the mobile phone shown in Figure 1 (a) considered in free space configuration. The outputs are then the two frequencies of the antenna for which the return losses are the minimum. The results for either one or two random variables are examined. The first random variable is the relative permittivity of the antenna support and the second random variable is the relative permittivity of the casing. Tables 1 and Table 2 show the comparison results between MC and UT for the cases of one and two random variables, respectively. The results obtained using UT correspond to 3 and 36 runs for the cases of one and two random variables, respectively. Several hundreds of runs are required when applying MC for either one or two random variables.

Tables 3 and 4 show the results of the uncertainty in the peak 10g SAR values for the two mobile phones for the right/cheek position against the SAM phantom using UT.

Figure 1. (a) Numerical model of the first dual-band commercially available mobile phones, (b) the second mobile phone model and (c) mobile phone placed against the right/cheek position of the SAM phantom.

Table 1. Mean and uncertainty calculated by MC and UT (Input: x_1).

Random variable	Method	GSM 900MHz(MHz)	DCS 1800MHz(MHz)
2.33±21.5%	MC	894.5±1.26%	1718.5±1.47%
	UT	896.1±1.29%	1718.8±1.51%

Table 2. Mean and uncertainty calculated by MC and UT (Inputs: x_1 and x_2).

Random variable	Method	GSM 900 MHz (MHz)	DCS 1800 MHz(MHz)
2.33±21.5%	MC	895.3±1.26%	1719.5±1.97%
1.8±22.2%	UT	896.1±1.29%	1719.3±2.51%

Table 3. Mean and uncertainty of the peak 10g SAR of moodel 1 obtained by UT.

Random variable	SAR10g (GSM)[W/kg]	SAR10g (DCS) [W/kg]
2.33±21.5%	1.35±2.58%	0.43±1.92%
$2.33\pm21.5\%, 1.8 \pm 22.2\%$	1.31±1.69%	0.45±2.58%

Table 4. Mean and uncertainty of the peak 10g SAR of moodel 2 obtained by UT.

Random variable	SAR10g (GSM)[W/kg]	SAR10g (DCS) [W/kg]
2.33±21.5%	1.07±2.80%	0.61±11.48%
$2.33\pm21.5\%, 3 \pm 13.3\%$	1.07±3.42%	0.67±9.97%

4. References

[1] Joint Committee for Guides in Metrology, Sept. 2008, Evaluation of measurement data Guide to the expression of uncertainty in measurement, 1st edition, *JCGM 100:2008.*

[2] P.J.Roache, January.1997, Quantification of uncertainty in computatinal fluid dynamics, *Annual Review of Fluid Mechanics*, vol. **29**, pp 123-160.

[3] L.Mathelin et al, March.2005, Stochastic approaches to uncertainty quantification in CFD simulations, *Numerical algorithms,springer*, vol. **38**, pp 209-236.

[4] IEEE, 2003, Recommended practice for determining the peak spatial-average Specific Absorption Rate (SAR) in the human head from wireless communications devices: measurement techniques, *IEEE Standard 1528.*

[5] IEEE/IEC62704-1, June. 2013, Determining the peak spatial-average specific absorption rate (SAR) in the human body from wireless communications devices, 30 MHz 6 GHz Part 1: General requirements for using the finite difference time-domain (FDTD) method for SAR calculations, Draft.

[6] IEC62704-3, June. 2013, Determining the peak spatial-average specific absorption rate (SAR) in the human body from wireless communications devices, 30 MHz 6 GHz Part 3: Specific requirements for using the finite difference time-domain (FDTD) method for SAR calculations of mobile phones, Draft.

[7] V. Monebhurrun et al, Feb. 2014, A benchmark CAD mobile phone model for the specific absorption rate calculation, *IEEE Transactions on Magnetics*, vol. **50**.

[8] S.J. Julier and J.K.Uhlmann, 1997,A consistent,debiased method for coverting between polar and cartesian coordinate systems. *Proc. AeroSense: Acquisition,Track ing and pointing XI*, vol. **3086**, pp 110-121.

[9] J. A.Bucklew, P. Ney, and J. S.Sadowsky, 1990 , Monte Carlo simulation and large deviations theory for uniformly recurrent Markov chains, *Journal of Applied Probability*, vol. **27**, pp 44-59.

[10] H.M.Menegaz et al, Dec.2011, A new smallest sigma set for the Unscented Transform and its applications on SLAM, *50th IEEE Conference on Decision and European Control Conference*

[11] L. R.A.X de Menezes, Mar. 2008, Efficient computation of stochastic electromagnetic problems using unscented transforms, *IET Science, Measurement and Technology*, vol. **2**, pp 88-95.

[12] M. Clemens, E. Gjonaj, P. Pinder, and T. Weiland , July.2000, Numerical Simulation of Coupled Transient Thermal and Electromagnetic Fields with the Finite Integration Method, *IEEE Transactions on Magnetics*, vol. **36** , pp 1448-1452.

Application of the SAROTA index in real-life scenario

A Rojatkar, V Monebhurrun
SUPELEC – Department of Electromagnetics, DRE-L2S
3 rue Joliot-Curie, 91192 Gif –sur-Yvette cedex, France

E-mail: ashish.rojatkar@supelec.fr

Abstract. A unique parameter referred to as the SAROTA index which accounts for both the specific absorption rate (SAR) and the over-the-air (OTA) performance of a mobile phone was previously proposed to characterize the real-life exposure. The applicability of the SAROTA index was confirmed using SAR and total radiated power (TRP) data obtained under laboratory conditions wherein the power control (PC) enforced on the mobile phone was implemented artificially. Herein the investigation is extended to measurements conducted for the speech mode of operation in real-life scenarios. Based on the actual PC implemented during the communication with the base station, the instantaneous and average real-life exposure experienced by the mobile phone user is analyzed and compared to the predicted SAROTA index. To capture the PC in real-time, a set of hardware modified phones with embedded network monitoring software are used. The instantaneous uplink transmit power level (TX_LEV) along with various downlink parameters such as the receive signal level (RX_LEV) and received signal quality (RX_QUAL) of the communication link are thus available for performing a comprehensive RF exposure analysis.

1. Introduction

International guidelines have proposed the peak spatial-average specific absorption rate (SAR) as the metric to quantify the limit of radio frequency (RF) exposure due to mobile phones [1]. The procedure to evaluate the SAR compliance of mobile phones is described in international measurement standards [2,3]. For example, in European countries, the peak 10 g average SAR value of a mobile phone operated at the ear should not exceed 2 W/kg. In some countries, such as in France, it is mandatory that the actual SAR value be clearly displayed as information for the general public [4]. According to the current test procedures, the unique 10 g average SAR value actually corresponds to the maximum value obtained for various test configurations using the specific anthropomorphic mannequin (SAM) head phantom. Measurements are performed while the mobile phone is operated at the maximum power level for four intended use positions –left/cheek, left/tilt, right/cheek, right/tilt– at the centre frequency of a given operating band, following which additional measurements are performed at the two extreme frequencies of the examined band for the configuration which yielded the maximum SAR value. The same procedure is applied to all the operating frequency bands of the mobile phone and the displayed SAR value is the maximum obtained from all the test configurations. For a quad-band mobile phone –Global System for Mobile (GSM) communication 900 MHz, GSM 1800 MHz, Universal Mobile Telecommunication System (UMTS) 2100 MHz and Long-Term Evolution (LTE) 2600 MHz– a total of 24 elementary measurements are required to obtain the maximum 10g SAR value.

In real-life scenarios, the power emitted by the mobile phone and the frequency band of operation will be essentially dictated by the network operator [5-7]. Depending on the network condition at a given moment and location e.g. received signal level (RX_LEV) and quality (RX_QUAL), network load etc., the mobile phone may shift from one frequency band to another. A power control (PC) mechanism is also deployed so that the power emitted by the mobile phone is optimized to ensure an acceptable quality of the communication. Clearly, the average real-life exposure experienced by the mobile phone user will be relatively lower than the maximum 10 g SAR value measured under laboratory condition. Furthermore, the average real-life exposure will not only depend on the maximum 10 g SAR value but also on the over-the-air (OTA) performance of the mobile phone [8]. Both SAR and OTA data are therefore required to characterize the real-life exposure and the SAROTA index was previously proposed to account for both these parameters [9]. The applicability of the SAROTA index was previously investigated using SAR and total radiated power (TRP) data and the PC mechanism was simulated using a base station emulator [10]. Herein the investigation of the SAROTA concept is extended to real-life scenarios wherein the power control mechanism is dictated by the actual base station of a given network operator.

2. SAROTA

The SAROTA index is simply expressed as the ratio between the peak 10 g average SAR (SAR_{max} W/kg) and the TRP_{max} (W) measured under laboratory condition for the same test configuration, i.e. same mobile phone position against the SAM head phantom, the same traffic channel and the same maximum transmit power level:

$$SAROTA = \left[\frac{SAR_{\max}}{TRP_{\max}} \right]_{Configuration} \quad (1)$$

The SAROTA index (kg^{-1}) provides a unique parameter to quantify the performance of a mobile phone in terms of both RF exposure and OTA performance. For example, it is predicted that a mobile phone with a high SAROTA index will lead to a relatively higher average real-life exposure than one with a low SAROTA index. It is also predicted that two mobile phones with the same peak 10 g SAR values but different TRP values will yield different SAROTA indices and consequently different average real-life exposures.

The average real life exposure (SAR_{real_life}) can be derived from the SAROTA index as follows:

$$SAR_{real\_life} = SAROTA \times TRP_{real\_life} \quad (2)$$

where TRP_{real_life} is the average TRP observed in real-life scenarios with power control mechanism enforced by the base station. It is deduced from the instantaneous power emitted by the mobile phone (TX_LEV).

3. Measurement Setup

Two sets of hardware modified mobile phones with embedded network monitoring software are employed for the study: set I consists of four samples of the same mobile phone model (labelled A1 to A4) and set II consists of three different mobile phone models (labelled M1 to M3). Since signal strength variations due to near vicinity object movement will be observed in a typical wireless network, the investigation is preferably undertaken in a closed indoor environment to minimize these variations and to gain insight of the PC mechanism experienced by the mobile phones. Set I is used to examine the stability of the results in this indoor environment which is an empty room containing only the measurement setup. Set II is then used to evaluate the real-life exposure due to the mobile phones placed against the right/cheek position of the SAM phantom as shown in figure 2. The measurements are performed for the GSM 900 MHz band with the mobile phones operating on three different networks (referred to as A, B and C) chosen such that the combination of the RX_LEV and RX_QUAL levels are distinct for each network at the given location. To account for the instantaneous signal strength variations, the measurement time is set to 30 minutes and the mobile phones are configured to play an audio file for the uplink to prevent activation of the discontinuous transmission (DTX) mode.

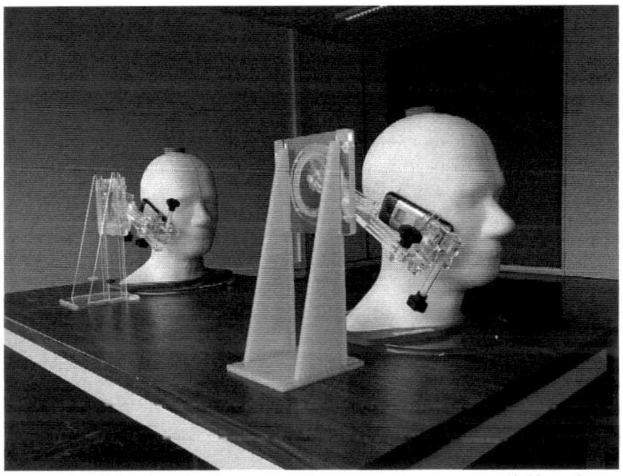

Figure 1. Experimental setup inside the indoor environment: one mobile phone calls another one, with both mobile phones placed against the right/cheek position of the SAM phantom.

4. Measurement Results

Figure 2(a) shows the results of the RX_LEV and TX_LEV obtained for one network operator using the four samples of set I. As expected, since the electromagnetic environment is rather similar at the location for a given network, all mobile phones show similar received and transmit characteristics. To demonstrate the impact of RX_LEV on the TX_LEV, the OTA performance of one mobile phone was modified by simply sticking a metallic strip over the antenna location (A2_mod). Figure 2(b) shows that the RX_LEV is then relatively lower and consequently the TX_LEV increases.

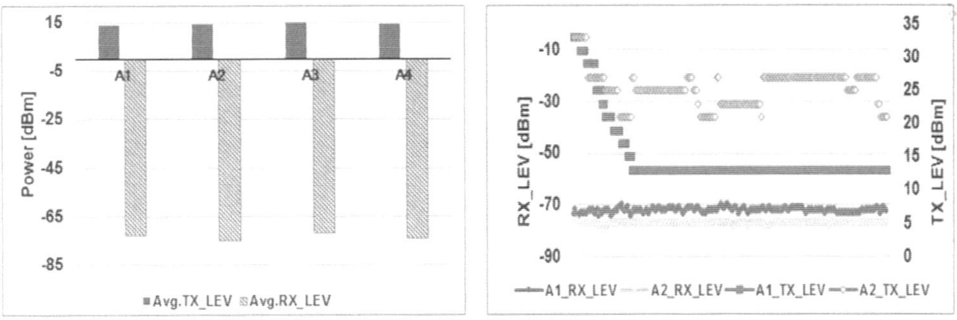

Figure 2. Measured TX and RX levels: (a) Set I (A1 to A4) (left), and (b) A1 and A2_mod.

Table 1 provides the results obtained using the three mobile phones from set II operating on the three different networks. Information about the propagation condition and the quality of a given network is gained from RX_LEV and RX_QUAL, respectively. TRP_{real_life} is representative of the overall network condition. For example, even when the received signal level is highest for network B (<-70 dBm) with relative bad quality (2-4), the mobile phones emit at higher power levels and the average real-life exposure is then much higher indicating channel specific behaviour. Furthermore, SAR_{real_life} values are consistent with the values predicted using the SAROTA index i.e. the lowest real-life exposure is observed for the mobile phone with the lowest SAROTA index (M1).

Table 1: $SAR_{real\ life}$ on three different networks at the given indoor location.

Network	Mobile Phone Model	SARmax (W/kg)	TRPmax (dBm)	SAROTA (kg^{-1})	Avg. RX_LEV (dBm)	TRPreal_life (dBm)	SARreal_life (W/kg)
A (RX_QUAL: 0)	M1	0.155	24.86	4.05	-74.5	7.76	0.024
	M2	0.299	24.52	8.44	-79.4	5.56	0.030
	M3	0.685	26.87	11.3	-77.5	8.39	0.078
B (RX_QUAL: 2-4)	M1	0.163	26.60	2.85	-67.4	20.48	0.32
	M2	0.299	26.77	5.03	-69.7	21.64	0.73
	M3	0.633	26.49	11.4	-67.1	21.54	1.62
C (RX_QUAL: 4-7)	M1	0.222	24.58	6.17	-81.2	10.84	0.075
	M2	0.292	23.64	10.1	-78.6	14.82	0.31
	M3	0.692	25.69	14.93	-78.7	15.60	0.068

* RX_QUAL range 0-7, with 0 = best receive quality and 7 = worst receive quality

5. Conclusion

The applicability of the SAROTA index as an indicator of the real-life exposure due to mobile phones was investigated. The effect of the power control mechanism was characterized for three different networks at a given indoor location. The real-life TRP provides information about the overall network condition including the effects of both received signal level and quality. The real-life exposure is consistent with the predicted SAROTA index. Similar evaluation will be performed for the outdoor scenarios in the future.

6. References

1. *ICNIRP guidelines for limiting exposure to time-varying electric, magnetic and electromagnetic fields (up to 300 GHz)*, Health Physics, 74 (4), 1998, pp. 494-522.
2. *Recommended practice for determining the peak spatial-average Specific Absorption Rate (SAR) in the human head from wireless communications devices: measurement techniques*, IEEE Standard 1528, 2013.
3. *Human exposure to radio frequency fields from hand-held and body-mounted wireless communication devices - Human models, instrumentation, and procedures - Part1: Procedure to determine the specific absorption rate (SAR) for hand-held devices used in close proximity to the ear (frequency range of 300 MHz to 3 GHz*, IEC Std 62209-1, 2005.
4. *Décret no 2010-1207 du 12 octobre 2010 relatif à l'affichage du débit d'absorption spécifique des équipements terminaux radioélectriques,* JORF no 0239 du 14 octobre 2010 page 18463, texte no 34.
5. N. T. Atanasov and G. L. Atanasova, An investigation of the effects from the user's hand and head over received level and adaptive power control of a GSM mobile phone is typical operating environment, The Environmentalist, Vol.27, Issue 4, Dec. 2007, pp. 585-591.
6. M. C. Gosselin, S. Kuehn and S. Kuster, Evaluation of the output power control of multi communication system mobile phones, in Proc. APSURSI, July 2010, pp. 1-4.
7. S. Kuhn and N. Kuster, "Field evaluation of the human exposure from multiband, multisystem mobile phones," IEEE Transactions on Electromagnetic Compatibility, Vol. 55 (2), Apr. 2013, pp. 275-287.
8. *Method of measurement for radiated RF power and receiver performance*, CTIA Certification Test Plan for Mobile Station Over The Air Performance, Rev. 3.0, 2009.
9. V. Monebhurrun, SAROTA: application of specific absorption rate (SAR) and over-the-air (OTA) data for the characterization of the real-life exposure due to mobile phones, IoP Conference Series: Materials Science and Engineering, 44, 012018, 2013.
10. A. Rojatkar and V. Monebhurrun, Experimental evaluation of the SAROTA concept for the evaluation of the real-life exposure due to mobile phones, BEMS 2013, Thessaloniki, 10-14 June 2013, 2 pages on CD.

Decentralized Gauss-Newton method for nonlinear least squares on wide area network

Lanchao Liu*, **Qing Ling**†**,and Zhu Han***

*Electrical and Computer Engineering Department, University of Houston, U.S.A
†Department of Automation, University of Science and Technology of China, China

E-mail: `lliu10@uh.edu`

Abstract. This paper presents a decentralized approach of Gauss-Newton (GN) method for nonlinear least squares (NLLS) on wide area network (WAN). In a multi-agent system, a centralized GN for NLLS requires the global GN Hessian matrix available at a central computing unit, which may incur large communication overhead. In the proposed decentralized alternative, each agent only needs local GN Hessian matrix to update iterates with the cooperation of neighbors. The detail formulation of decentralized NLLS on WAN is given, and the iteration at each agent is defined. The convergence property of the decentralized approach is analyzed, and numerical results validate the effectiveness of the proposed algorithm.

1. Introduction

The significant importance of the nonlinear least squares (NLLS) in applications of state estimation in power system [1], signal detection in wireless networks [2], and target tracking in mobile networks [3] has been appreciated for decades. The Gauss-Newton (GN) method, which can be seen as a modification of the Newton's method, is widely used to solve the NLLS [4]. The GN method finds the minimizer of the NLLS in an iterative fashion, and obtains the solution with provable local optimality and convergence rate. In this work, a decentralized GN method for NLLS on WAN is presented. In particular, only local GN Hessian matrix is used and limited communication is performed between neighboring agents. The decentralized optimization enjoys the advantage of scalability to network size, robustness to dynamic topologies and privacy preservation in data-sensitive applications [5–8]. A detailed formulation of the decentralized optimization problem for NLLS on WAN is provided, and the updating rule at each agent is explicitly given. Finally, we investigate the convergence property of the proposed algorithm, which turns out the convergence rate is related to the number of agents as well as the minimum node degree in the network. Numerical tests validate the performance of the proposed algorithm.

The contributions of this work are threefold. Firstly, we do not assume any specific structure for the global Hessian matrix, and proposed a decentralized GN method for NLLS use only local Hessian matrix. Whereas the localization application in [3] has a block-wise Jacobian matrix which is convenient to decompose, and needs global Hessian through network-wide consensus. [1] proposes a generalized gossip-based GN method, which still requires global Hessian through Gossip exchange. Secondly, we proved the local superlinearly convergence property of the proposed algorithm. Finally, we validated the proposed method through numerical simulations.

2. Problem Formulation

2.1. Centralized Nonlinear Least Squares

Consider an unknown variable $\tilde{\mathbf{x}} \in \mathbb{R}^n$ in a network, and m observations are obtained through a vector-valued function $\mathbf{h}(\tilde{\mathbf{x}}) = (h_1(\tilde{\mathbf{x}}), \ldots, h_m(\tilde{\mathbf{x}})) : \mathbb{R}^n \to \mathbb{R}^m$. Each entry in function $\mathbf{h}(\tilde{\mathbf{x}})$ is a real value function and not necessarily convex. Let $\mathbf{z} \in \mathbb{R}^m$ denote the observations as $\mathbf{z} = \mathbf{h}(\tilde{\mathbf{x}}) + \mathbf{e}$, where $\mathbf{e}$ stands for measurement errors, which are assumed zero mean and known covariance $\mathbf{R} \in \mathbb{R}^{m \times m}$. The unknown variable $\tilde{\mathbf{x}} \in \mathbb{R}^n$ can be estimated by the NLLS as follows:

$$\underset{\tilde{\mathbf{x}}}{\text{minimize}} \quad (\mathbf{z} - \mathbf{h}(\tilde{\mathbf{x}}))^\top \mathbf{R}^{-1} (\mathbf{z} - \mathbf{h}(\tilde{\mathbf{x}})). \tag{1}$$

The GN method can be adopted to solve (1) given that all the observations and functions are available at a central computing node. Specifically, define $\mathbf{r}(\tilde{\mathbf{x}}) = \mathbf{R}^{-1/2}(\mathbf{z} - \mathbf{h}(\tilde{\mathbf{x}}))$ and its Jacobian $\mathbf{J}(\tilde{\mathbf{x}}) = \partial \mathbf{r}(\tilde{\mathbf{x}}) / \partial \tilde{\mathbf{x}}$. Let $F(\tilde{\mathbf{x}}) = \|r(\tilde{\mathbf{x}})\|^2$, and problem (1) can be solved iteratively as $\tilde{\mathbf{x}}^{k+1} = \tilde{\mathbf{x}}^k + \alpha^k \mathbf{d}^k$, where the descent direction $\mathbf{d}^k$ at each iteration can be obtained by solving

$$\mathbf{J}^\top(\tilde{\mathbf{x}}^k)\mathbf{J}(\tilde{\mathbf{x}}^k)\mathbf{d} = \mathbf{J}^\top(\tilde{\mathbf{x}}^k)\mathbf{r}(\tilde{\mathbf{x}}^k). \tag{2}$$

The GN method can solve the problem (1) at a superlinear convergence rate with order at least two under Assumption 1. Note that majority of NLLS problems are non-convex, and in this paper we only consider the local convergence property of the algorithm. We assume Assumption 1 holds throughout the paper.

Assumption 1. *Consider a function $F(\tilde{\mathbf{x}})$, suppose the following assumptions hold.*

(i) The function $F(\tilde{\mathbf{x}})$ is continuous, differentiable and bounded below.
(ii) There exists a vector $\tilde{\mathbf{x}}^*$ such that the greatest lower bound can be achieved.
(iii) For $\delta > 0$, let S_δ denote the sphere $\{\tilde{\mathbf{x}} \mid \|\tilde{\mathbf{x}} - \tilde{\mathbf{x}}^*\|^2 \leq \delta\}$. The Hessian matrix $\mathbf{J}^\top(\tilde{\mathbf{x}})\mathbf{J}(\tilde{\mathbf{x}})$ is invertible in the sphere S_δ. For some $L > 0$, $M > 0$, $\delta > 0$, and for all $\tilde{\mathbf{x}}$ and $\tilde{\mathbf{y}}$ in S_δ, we have $\|\mathbf{J}^\top(\tilde{\mathbf{x}})\mathbf{J}(\tilde{\mathbf{x}}) - \mathbf{J}^\top(\tilde{\mathbf{y}})\mathbf{J}(\tilde{\mathbf{y}})\| \leq L\|\tilde{\mathbf{x}} - \tilde{\mathbf{y}}\|$ and $\sigma_{\min}(\mathbf{J}^\top(\tilde{\mathbf{x}})\mathbf{J}(\tilde{\mathbf{x}})) \geq \frac{1}{M} \geq 0$.

2.2. Decentralized Nonlinear Least Squares

In a multi-agent system consisted of N networked agents, each agent is engaged in its own monitoring and controlling task in the network, and at the same time, each agent is cooperating with other agents in the context of estimating the global system states $\tilde{\mathbf{x}}$. Suppose these N agents are loosely coupled; there is very little, if any, central coordination and control among those agents, and each agent is able to exchange information with its neighbors. The system states $\tilde{\mathbf{x}}$ can be obtained by solving the following optimization problem:

$$\begin{aligned} \underset{\mathbf{x}_i \ldots \mathbf{x}_N}{\text{minimize}} \quad & f(\mathbf{x}) = \sum_{i=1}^{N} (\mathbf{z}_i - \mathbf{h}_i(\mathbf{x}_i))^\top \mathbf{R}_i^{-1} (\mathbf{z}_i - \mathbf{h}_i(\mathbf{x}_i)), \\ \text{subject to} \quad & \mathbf{x}_i = \ldots = \mathbf{x}_N, \end{aligned} \tag{3}$$

where $\mathbf{z}_i$ is the local observation which is a subset of $\mathbf{z}$, i.e., $\mathbf{z} = (\mathbf{z}_1; \ldots; \mathbf{z}_N)$, and $\mathbf{h}_i$ is the local observation function which is a subset of $\mathbf{h}$, i.e., $\mathbf{h} = (\mathbf{h}_1; \ldots; \mathbf{h}_N)$. $\mathbf{R}_i^{-1}$ is the covariance matrix of local noise vector $\mathbf{e}_i$. $\mathbf{x}_i \in \mathbb{R}^n$ is the local duplicate of $\tilde{\mathbf{x}}$, and $\mathbf{x} = (\mathbf{x}_1; \ldots; \mathbf{x}_N) \in \mathbb{R}^{Nn}$.

3. Proposed Algorithms

For concreteness, the network model of agents is first described. Specifically, consider a graph $\mathcal{G} = (\mathcal{V}, \mathcal{E})$. $\mathcal{V}$ represents the set of agents, and $\mathcal{E}$ represents the set of communication links between each pair of agents. An arc e is associated with an order pair (i, j) as $e \sim (i, j)$, which means the information is transmitted from agent i to agent j. Assume the graph formed by the agents is connected. By introducing the auxiliary variables $\mathbf{w}_{ij} \in \mathbb{R}^n$ associated with each arc $e \sim (i,j) \in \mathcal{E}$, problem (3) can be reformulated as:

$$\underset{\{\mathbf{x}_1,\ldots,\mathbf{x}_N\}}{\text{minimize}} \quad \sum_{i=1}^{N} (\mathbf{z}_i - \mathbf{h}_i(\mathbf{x}_i))^\top \mathbf{R}_i^{-1} (\mathbf{z}_i - \mathbf{h}_i(\mathbf{x}_i)), \tag{4}$$

$$\text{subject to} \quad \mathbf{x}_i = \mathbf{w}_{ij}, \quad \mathbf{x}_j = \mathbf{w}_{ij}, \quad \forall (i,j) \in \mathcal{E}, \tag{5}$$

where $\mathbf{w}_{ij}$ is used to enforce the equality of variables $\mathbf{x}_i$ and $\mathbf{x}_j$ for agents i and j connected by arc (i,j). We use compact notations in the following for the sake of discussion simplicity. Concatenating $\mathbf{w}_{ij}$ in vector $\mathbf{w}$, problem (4) can be reformulated as:

$$\underset{\mathbf{x},\mathbf{w}}{\text{minimize}} \quad f(\mathbf{x}), \quad \text{subject to} \quad \mathbf{A}_s\mathbf{x} - \mathbf{w} = 0, \quad \mathbf{A}_d\mathbf{x} - \mathbf{w} = 0, \tag{6}$$

where $\mathbf{A}_s$ and $\mathbf{A}_d$ are the extended arc source matrix and extended arc destination matrix for the network graph $\mathcal{G}$, respectively. Stacking $\mathbf{A}_s$ and $\mathbf{A}_d$ to form $\mathbf{A} = [\mathbf{A}_s; \mathbf{A}_d] \in \mathbb{R}^{2Mn \times Nn}$. The optimization problem (6) reduces to:

$$\underset{\mathbf{x},\mathbf{w}}{\text{minimize}} \quad f(\mathbf{x}), \quad \text{subject to} \quad \mathbf{A}\mathbf{x} + \mathbf{B}\mathbf{w} = 0, \tag{7}$$

where $\mathbf{B} = [-\mathbf{I}_{Mn}; -\mathbf{I}_{Mn}] \in \mathbb{R}^{2Mn \times Mn}$. The GN method is utilized to solve the optimization problem (7), where the updates of $\mathbf{x}$ are implemented in a decentralized fashion. The local update rule at each agent is given in the following proposition.

Proposition 1. *Consider iterates $\mathbf{x}^k$ and $\mathbf{z}^k$ with the initialization $\mathbf{E}_u\mathbf{x}^0 = 2\mathbf{w}^0$, the iterates $\mathbf{x}_i^k$ at each agent i can be iteratively generated by the following recursions for $k > 0$:*

$$\mathbf{x}_i^{k+1} = \mathbf{x}_i^k + \alpha_i^k \mathbf{d}_i^k, \tag{8}$$

where α_i^k is a positive constant, and $\mathbf{d}_i^k$ is the descent direction which can be determined as

$$(\mathbf{J}_i^\top(\mathbf{x}_i^k)\mathbf{J}_i(\mathbf{x}_i^k) + \rho\nu_i\mathbf{I}_n)\mathbf{d}_i^k = \mathbf{J}_i^\top(\mathbf{x}_i^k)\mathbf{r}_i(\mathbf{x}_i^k) + \rho[\nu_i\mathbf{x}_i - \frac{1}{2}\sum_{j\in\mathcal{N}_i}(\mathbf{x}_i^{k-1} + \mathbf{x}_j^{k-1})],$$

where $\mathbf{r}_i(\mathbf{x}_i) = \mathbf{R}_i^{-1/2}(\mathbf{z}_i - \mathbf{h}_i(\mathbf{x}_i))$ and $\mathbf{J}_i(\mathbf{x}_i)$ is its Jacobian $\mathbf{J}_i(\mathbf{x}_i) = \partial\mathbf{r}_i(\mathbf{x}_i)/\partial\mathbf{x}_i$. ν_i is the degree of agent i in the network, and $\mathcal{N}_i$ denotes the neighbors of agent i.

Proof. The proof is omitted here due to the space limitation. □

The local convergence property and convergence rate of the proposed decentralized approach are given by the following theorem.

Theorem 1. *Suppose the Assumption 1 holds, and the start point of each agent $\mathbf{x}_i^0$ is in S_δ. The sequence $\{\mathbf{x}^k\}$ generated by the update rule given in Proposition 1 is defined, and converges to $\mathbf{x}^* = \{\tilde{\mathbf{x}}^*; \ldots; \tilde{\mathbf{x}}^*\}$. Furthermore, we have*

$$\|\mathbf{x}^{k+1} - \mathbf{x}^*\| \le \frac{ML\sqrt{N}}{2(1 + M\rho\ max(\nu_i))}\|\mathbf{x}^k - \mathbf{x}^*\|^2, \tag{9}$$

where N is the number of agents and $\max(\nu_i)$ is the maximum node degree in the network.

Proof. The proof is omitted here due to the space limitation. □

Remark that Theorem. 1 illustrates the local convergence property of the proposed decentralized approach, which converges to the optimal solution superlinearly. The convergence rate is related to the number of agents as well as the minimum node degree in the network.

4. Numerical Results

A bidirectionally connected ring network composed of $N = 100$ agents is considered here, in which each agent connects to exactly two other agents. The unknown system states in the network is $\tilde{\mathbf{x}} \in \mathbb{R}^3$. The observation function $\mathbf{h}_i(\mathbf{x}_i)$ at each agent i is defined as:

$$\mathbf{h}_i(\mathbf{x}_i) = a_i(\mathbf{x}_i(1)^2 + \mathbf{x}_i(2)^2) + b_i\mathbf{x}_i(2)\sin(\mathbf{x}_i(2) - \mathbf{x}_i(3)) + c_i\mathbf{x}_i(1)\mathbf{x}_i(2), \tag{10}$$

where a_i, b_i, and c_i are i.i.d. random variables follow the standard normal distribution. It is seen that observation function $\mathbf{h}_i(\mathbf{x}_i)$ is a nonlinear function with a quadratic term, a trigonometric term and a cross product term. The agents in the network work cooperatively to estimate the unknown system states $\tilde{\mathbf{x}}$ in a decentralized fashion. The convergence result is depicted in Fig. 1. It is shown that the proposed algorithm is effective in the sense that after a moderate number of iterations, the iterates converge to the optimal values. To investigate the performance of the proposed decentralized approach at each agent, the root-mean-square error (RMSE) of the estimate at each agent is calculated. The best RMSE (agent 2), the worst RMSE(agent 52) and the average RMSE are described in Fig. 2. It can be seen that at each agent, the RMSE decreases as the iteration increases. Furthermore, the convergence rate at each agent is different.

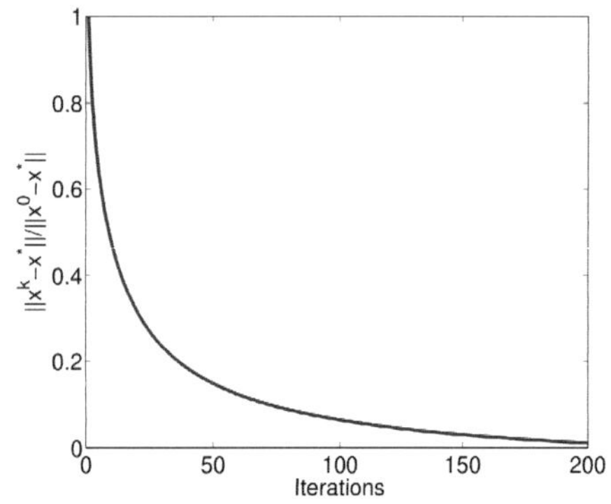

Figure 1. Convergence performance of the proposed algorithm.

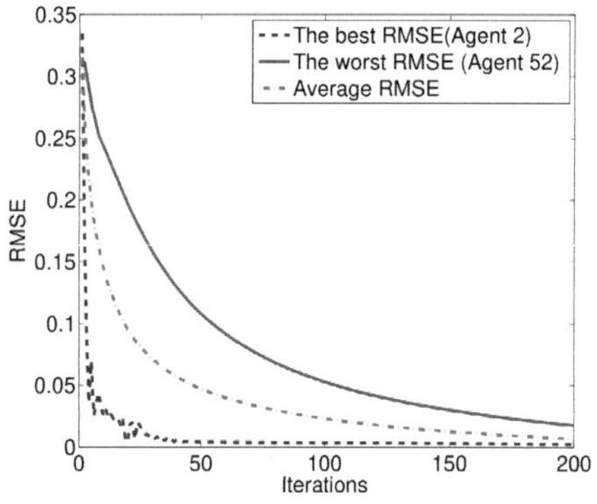

Figure 2. RMSE performance of the proposed algorithm.

5. Conclusions

We have presented a decentralized approach of GN method for NLLS on WAN. We have given the updating rule at each agent explicitly, and investigated the local convergence of the proposed algorithm. Numerical simulations validated the effectiveness of the proposed algorithm.

References

[1] Li X and Scaglione A 2013 *IEEE Transactions on Signal Processing* **61** 5231–5246
[2] Stoica P, Mose R, Friedlander B and Soderstrom T 1989 *IEEE Transaction on Acoustics, Speech and Signal Processing* **37** 378–392
[3] Bejar B, Belanovic P and Zazo S 2010 *2010 Conference Record of the Forty Fourth Asilomar Conference on Signals, Systems and Computers (ASILOMAR)* (Pacific Grove, CA)
[4] Bertsekas D P 1999 *Nonlinear Programming* (Nashua, USA: Athena Scientific)
[5] Bertsekas D P and Tsitsiklis J N 1989 *Parallel and Distributed Computing: Numerical Methods* (NJ, USA: Prentice-Hall)
[6] Nedic A and Ozdaglar A 2009 *IEEE Transaction on Automatic Control* **54** 48–61
[7] Ram S, Nedic A and Veeravalli V 2010 *Journal of Optimization Theory and Applications* **147** 516–545
[8] Srivastava K and Nedic A 2011 *IEEE Journal of Selected Topics in Signal Processing* **5** 772–790

2nd Radio and Antenna Days of the Indian Ocean (RADIO 2014) IOP Publishing
IOP Conf. Series: Materials Science and Engineering **67** (2014) 012022 doi:10.1088/1757-899X/67/1/012022

An Improved Call Admission Control Mechanism with Prioritized Handoff Queuing Scheme for BWA Networks

Prasun Chowdhury, Iti Saha Misra
Department of Electronics and Telecommunication Engineering
Jadavpur University, Kolkata - 700032, India

E-mail: prasun.jucal@gmail.com

Abstract. Nowadays, due to increased demand for using the Broadband Wireless Access (BWA) networks in a satisfactory manner a promised Quality of Service (QoS) is required to manage the seamless transmission of the heterogeneous handoff calls. To this end, this paper proposes an improved Call Admission Control (CAC) mechanism with prioritized handoff queuing scheme that aims to reduce dropping probability of handoff calls. Handoff calls are queued when no bandwidth is available even after the allowable bandwidth degradation of the ongoing calls and get admitted into the network when an ongoing call is terminated with a higher priority than the newly originated call. An analytical Markov model for the proposed CAC mechanism is developed to analyze various performance parameters. Analytical results show that our proposed CAC with handoff queuing scheme prioritizes the handoff calls effectively and reduces dropping probability of the system by 78.57% for real-time traffic without degrading the number of failed new call attempts. This results in the increased bandwidth utilization of the network.

1. Introduction

In recent years, various BWA networks such as the Worldwide Interoperability for Microwave Access (WiMAX) [1] or the Long Term Evolution (LTE) [2] are being developed rapidly to support heterogeneous applications such as real-time video, voice over IP (VoIP) and web browsing. To manage the heterogeneous incoming calls that have stringent QoS requirements more sophisticated CAC mechanism are to be developed [3]. Moreover, to control handoff dropping probability certain measures have to be employed to prioritize handoff calls over new calls since it is commonly accepted that dropping an ongoing handoff call is more annoying than dropping a new incoming call [4]. Several reservation and non reservation based policy with bandwidth degradation CAC mechanism [5 - 7] have been proposed in the literature to mitigate the problem related to the handoff calls. However, the concept of queuing to hold handoff request when bandwidth is unavailable even after the bandwidth degradation of admitted calls has not been included in the more efficient non reservation based CAC mechanism [7]. That is why in this paper, we have proposed an improved CAC mechanism that allow queuing of handoff call attempts if necessary to reduce dropping probability to a further extent and thereby increasing the user satisfaction in terms of seamless communication. An analytical model based on Continuous Time Markov Chain (CTMC) [8] is developed for the more realistic analysis of the proposed CAC mechanism. BWA services are categorized into Class-1, Class-2 and Class-3 traffic depending upon their QoS requirements and tasks to perform. The services which come under the Class-1 and Class-2 traffic are real-time (RT) VoIP and Video respectively while Class-3 traffic performs tasks related to non real-time (NRT) FTP, e-mail and web browsing.

2. Description of the proposed CAC mechanism

A BWA system with homogenous cells is considered along with a single Base Station (BS) under a reference cell. The proposed CAC mechanism is shown in figure 1. Both newly originated and handoff calls are admitted in the network if bandwidth is available. But if bandwidth is not available, the new call is admitted by allowable bandwidth degradation of the ongoing Class-3 calls and the handoff calls are queued and get admitted when an ongoing call is terminated with a higher priority than the new call. Allowable bandwidth degradation of both Class-2 and Class-3 ongoing calls in the target cell is executed to admit the handoff calls. Three separate queues with queue size QL_1, QL_2 and QL_3 are used to hold the handoff request of three different traffic classes respectively. Priority (Class-1>Class-2>Class-3) is maintained while admitting the handoff calls. The arrival process of the Class-1, Class-2 and Class-3 new calls is Poisson with average rates of λ_1, λ_2 and λ_3 respectively and that of for handoff calls λ_{h1}, λ_{h2} and λ_{h3} respectively.

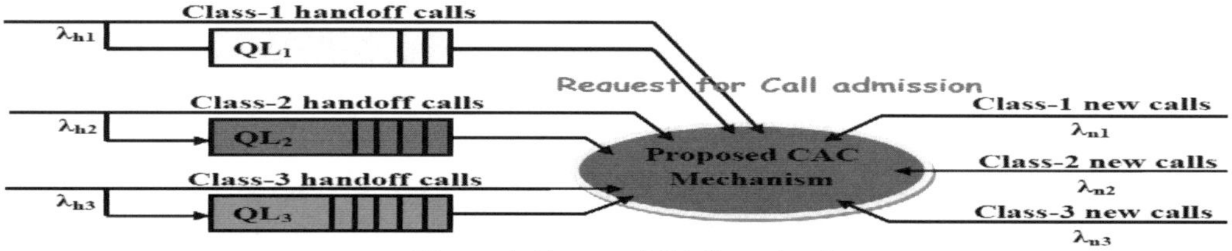

Figure 1. Proposed CAC mechanism

3. Analytical model for proposed CAC mechanism

It is assumed that BS changes state from one to another upon the admission or rejection or queuing of a traffic call. Further, the BS either admits or rejects or queued up only one traffic call at a particular instant of time.

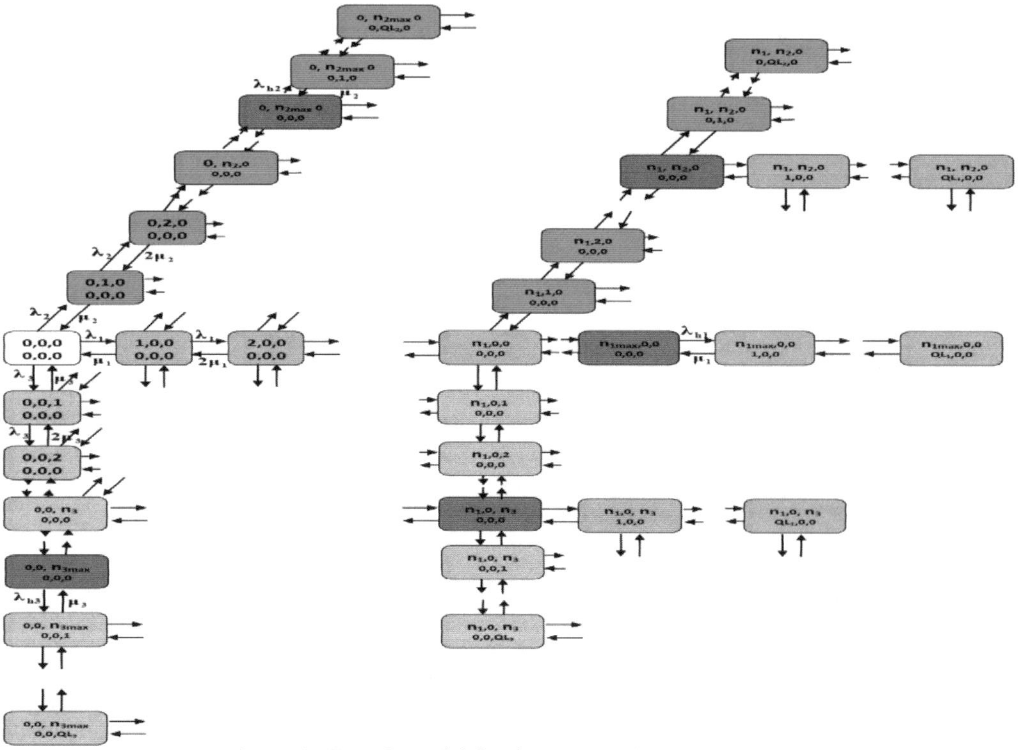

Figure 2. CTMC Model for the proposed CAC mechanism

So, the next state of the BS depends only on the present state of the BS but does not depend on the previous states of the BS. Therefore, the states of the BS form a Markov Chain. In this scenario, the BS can uniquely be represented in the form of a six dimensional Markov Chain based on the number of admitted connections (n_1, n_2, n_3) in the network as well as number of handoff connections (h_1, h_2, h_3) waiting in the queue of each traffic type. The state space $\begin{bmatrix} n_1, n_2, n_3 \\ h_1, h_2, h_3 \end{bmatrix}$ for our proposed CAC scheme is obtained based on the Markov chain in figure 2.

4. Results and Discussions

The values of the system parameters have been set as shown in table 1 for the performance evaluation. From the CTMC model steady state probabilities of each state of the BS have been evaluated and are used to derive several QoS performance parameters such as bandwidth utilization (BU), handoff queuing latency, new connection blocking probability (NCBP) and handoff connection dropping probability (HCDP).

Table 1. Values of the system parameters.

Parameters	Values
Arrival rate ratio of Class-1, 2, 3 traffic both for ($\lambda_1 : \lambda_2 : \lambda_3$) and ($\lambda_{h1} : \lambda_{h2} : \lambda_{h3}$)	3:2:1
Mean service time	$1/\mu_1 = 1/\mu_2 = 1/\mu_3$
Specified Handoff latency (SHL)	< 50 ms
Total BW in kbps	15360 kbps
Queue Length (QL_i)	Infinite
BW for Class-1(B_1) in kbps	256
BW for Class-2(B_2) in kbps	1024 (max), 512 (min)
BW for Class-3(B_3) in kbps	1024 (max), 256 (min)

Figure 3a shows the handoff queuing latency against handoff call arrival rate (λ_h) for all traffic classes. With increase in λ_h, mean number of handoff calls waiting in the queue increase due to unavailability of bandwidth in the network and thereby increasing mean queuing latency of each traffic class. Observations reveal that handoff queuing latency violates specified handoff latency (<50ms) whenever λ_h exceeds 8.5 and 3.5 respectively for Class-2 and Class-3 traffic. For highest priority delay sensitive Class-1 traffic, handoff queuing delay is observed to be negligible. Figure 3b,c,d show the handoff call dropping probability against λ_h of Class-1, Class-2 and Class-3 traffic respectively with handoff queuing scheme and without handoff queuing scheme. A significant improvement in the dropping probability of Class-1 RT traffic is observed when the CAC mechanism is incorporated with the handoff queuing scheme. The reduction in the dropping probability of Class-1 traffic is observed to be 78.57% at arrival rate =10. Dropping probability is also observed to be reduced for Class-2 RT traffic (figure 3c) with the introduction of the queuing scheme in the CAC mechanism. But close observation reveals that the dropping probability of Class-2 traffic does not improve rather remains same whenever its call arrival rate crosses 8.5. This is because of the violation of specified handoff latency above this arrival rate. On the other hand, from figure 3d it has been observed that dropping probability of Class-3 traffic is not quiet impressive. Improvement in the performance occurs only up to the arrival rate = 3.5 and beyond this the performance is observed to be diminished. This happens because of the two main reasons, Firstly, the violation of the specified handoff latency and secondly, assignment of the lowest priority among all other services. However, Class-3 traffic being NRT traffic its speculative performance does not hamper the overall performance of the proposed CAC mechanism in the network. The proposed CAC mechanism with handoff queuing scheme enhances the BU of the

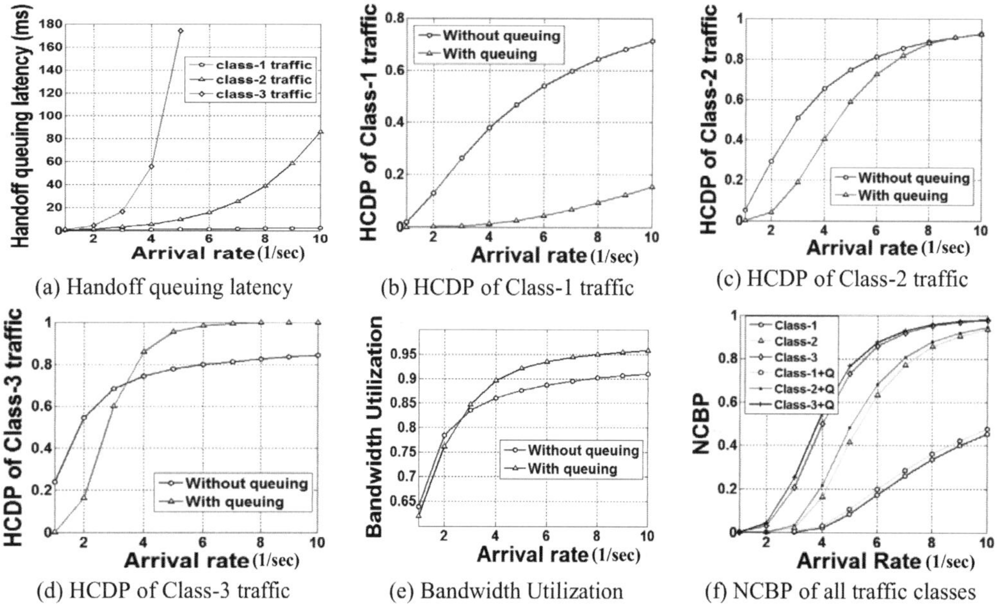

(a) Handoff queuing latency (b) HCDP of Class-1 traffic (c) HCDP of Class-2 traffic

(d) HCDP of Class-3 traffic (e) Bandwidth Utilization (f) NCBP of all traffic classes

Figure 3. Results for performance evaluation

network by 5 % as observed from figure 3e. Moreover, the NCBP for Class-1, Class-2 and Class-3 traffic both for 'with queuing scheme (+Q)' and 'without queuing scheme' as observed in figure 3f are more or less same. We have only found up to 3% deviation in the curve by introducing the handoff queuing scheme. It reveals that introduction of the handoff queuing scheme does not increase the NCBP of the system. Hence, in other words it can be said that QoS of the BWA network is improved in terms of HCDP and BU without degrading NCBP by our proposed CAC mechanism.

5. Conclusion

In this paper, we propose an improved CAC mechanism with prioritized handoff queuing scheme that handles the handoff calls whenever no bandwidth is available even after the bandwidth degradation of the admitted calls. Handoff calls, if any waiting in the queue get admitted when an ongoing call is terminated with a higher priority than a newly incoming call. An analytical model using CTMC is developed and evaluated for the proposed scheme to estimate the system performance. It is seen that our proposed scheme can achieve the dropping probability for the different traffic classes at a desired level without much degrading the blocking probability and increase the user satisfaction in terms of seamless communication while at the same time enhances resource utilization.

References

[1] IEEE standard for local and metropolitan area networks- part 16: 2006 *Air interface for fixed and mobile broadband wireless access systems, IEEE 802.16e-2005*

[2] Ghosh A, Ratasuk R, Mondal B, Mangalvedhe N and Thomas T 2010 *IEEE Wireless Communications* **17** 10-22

[3] Ray S.K, Pawlikowski K and Sirisena H 2010 *IEEE Comm. Surveys & Tutorials* **12** 376-399

[4] Sgora A and Vergados D.D 2009 *IEEE Comm. Surveys & Tutorials* **11** 57-77

[5] Hou F, Ho P and Shen X 2006 *IEEE GLOBECOM* 1-5

[6] Ge Y and Kuo G.S 2006 *IEEE 64th Vehicular Technology Conf.* 1-5

[7] Kalikivayi S, Misra I. S and Saha K 2008 *IEEE GLOBECOM* 1-6

[8] Ross S. M 2001 Probability Models for Computer Science (*Elseveir*)

IOP Conf. Series: Materials Science and Engineering **67** (2014) 012023 doi:10.1088/1757-899X/67/1/012023

Precoder selection scheme based on message-passing approach: a practical perspective

Igor M Guerreiro[1], Charles C Cavalcante[1] and Dennis Hui[2]

[1] Wireless Telecomm. Research Group, Universidade Federal do Ceará, Fortaleza, Brazil
[2] Ericsson Research, San José, California, USA

E-mail: {igor,charles}@gtel.ufc.br

E-mail: dennis.hui@ericsson.com

Abstract. This work addresses a distributed technique to precoding matrices coordination in a multi-cell network. We propose an iterative method based on a message-passing algorithm in factor graphs. Finite precoder codebooks for different MIMO setups are adopted for beam forming purposes. Conversely, practical cellular networks usually solve the precoder selection problem in a non-iterative manner based on the best response method. Nevertheless, this method may not provide a near-optimal solution. For more realistic results, the wireless channel is modeled based on measured data that comprises spatial correlation effects. Evaluations on the potential of such an approach are provided and its performance is compared with the selfish/greedy approach. The capability of reaching the globally optimal solution is evaluated, as well as its performance per iteration. Simulation results for the precoder selection example are presented and discussed, which show that the graph-based technique generally obtains gain in system capacity over both the non-iterative and selfish approaches. Besides, the proposed method usually reaches the global optima in an efficient manner in terms of signaling load.

1. Introduction

In a cellular network, each cell usually needs to set a parameter value, such as transmit power and beam direction, preferably in a compatible way with the settings of the neighboring cells, to achieve a certain notion of optimality, e.g. maximizing system throughput [1, 2]. The choice made by one cell on a local parameter often affects the interference level experienced by its immediate neighbors and hence their respective choices made on their local parameters, which in turn would influence the choices made by their neighbors' neighbors. In some cases, the parameter is dynamic and requires coordination to be continually performed. Therefore, a systematic methodology for coordinating the choices of any parameters across the network is desired. Moreover, to facilitate flexible, dense deployment of small base-stations in future cellular networks, there is an increased interest in methods of performing the coordination of parameters among neighboring cells in an autonomous and distributed fashion without a central controller, as any unplanned addition (or removal) of base-stations can substantially alter the system topology and thus the preferred settings.

In this paper, we propose the use of the min-sum algorithm [3] to precoder selection in a distributed way. Different from our work in [4], a discrete set of complex weighting precoding matrices is considered to be coordinated. Such an algorithm can then be applied in order for all nodes, through iterative message exchange with their respective neighboring nodes, to decide upon the best set of precoders that can collectively maximize the system capacity. It allows each communication node to be indecisive of its own decision until sufficient information is accumulated. The performance of the graph-based method against a greedy approach for coordination of discrete parameters in a wireless communication network is evaluated.

2. Problem Formulation

Consider a communication network with N communication nodes in a multi-cell multiple-input-multiple-output (MIMO) system. A communication node represents a pair of base-station (BS) and its associated user equipment (UE), each one in a cell, where only downlink transmissions are considered. Each BS has N_t available transmit antennas and each UE has N_r receive antennas. Let p_i denote a discrete parameter of the ith communication node, or simply node i, whose value is drawn from a finite set $\mathcal{P}_i$ of $|\mathcal{P}_i|$ possible parameter values for that node, where $|\mathcal{P}_i|$ denotes the cardinality of $\mathcal{P}_i$, and let $\mathbf{p} \equiv \begin{bmatrix} p_1 & p_2 & \cdots & p_N \end{bmatrix}^T$ be a vector collecting all the parameters in the network, where $p_i \in \mathcal{P}_i$, $i = 1, 2, \ldots, N$. Each parameter p_i represent a precoding matrix index (PMI) for BS i indicating which precoder from a predetermined codebook that BS i should use at a certain radio resource block to transmit signals.

Each node i is associated with a list $\mathcal{N}_i$ of proper neighbor nodes (excluding node i). Also let $\mathcal{A}_i \equiv \mathcal{N}_i \cup \{i\}$ denote the neighbor list of node i. Let $\mathbf{p}_{\mathcal{A}_i}$ denote the vector of those parameters of nodes in $\mathcal{A}_i$. The BS i transmits vector $\mathbf{x}_i$ to its associated UE i. The vector $\mathbf{x}_i$ is defined as $\mathbf{x}_i = \sqrt{1/N_s}\mathbf{W}_i\mathbf{s}_i$, where N_s is the number of data streams, $\mathbf{s}_i$ is the $N_s \times 1$ spatially multiplexed (SM) symbol vector and $\mathbf{W}_i \in \mathcal{W}$ is the $N_t \times N_s$ precoding matrix specified by the parameter p_i. Here, $\mathcal{W}$ is the finite set of all precoding matrices available for every communication node in the network. In order to index the elements of $\mathcal{W}$, assume an index set $\mathcal{I}$, which is equivalent to set $\mathcal{P}_i$. Due to the lack of space, the matrix formulation of the incoming signal is omitted (please refer to the formulation in [4]).

Our goal is for each node i to find its own optimal parameter p_i^*, which is the corresponding component of the optimal global parameter vector $\mathbf{p}^*$ that minimizes the global metric given by

$$M(\mathbf{p}) \equiv \sum_{i=1}^{N} M_i(\mathbf{p}_{\mathcal{A}_i}), \tag{1}$$

where $M_i(\mathbf{p}_{\mathcal{A}_i})$ is the local metric of node i and represents the negative of the data throughput [5, 6] of the cell corresponding to BS i. It is given by $M_i(\mathbf{p}_{\mathcal{A}_i}) = -\log\det\left(\mathbf{I} + |g_{ii}|\mathbf{R}_i^{-1}\mathbf{H}_{ii}\mathbf{W}_i\mathbf{W}_i^H\mathbf{H}_{ii}^H\right)$, where $\mathbf{H}_{ii}$ and g_{ii} denote the channel and the path gain between BS i and UE served by BS i, respectively, and $\mathbf{R}_i$ denotes the covariance matrix of the noise-plus-interference experienced by the UE served by BS i.

This problem can be solved through joint optimization, which yields the optimal global parameter vector. A major issue of this approach is its huge computational complexity for large network size. Alternatively, distributed approaches can be adopted, which on one hand often provide sub-optimal solutions through greedy techniques such as non-cooperative games, but on the other hand can provide near-optimal solutions by using message pass in factor graph. In this case, the min-sum algorithm, can then be executed on a factor graph [7, 4], which simply iterates computing and passing messages between nodes. The description of factor graph and the formulation of the algorithm are omitted due to lack of space (please refer to [4]).

3. Analysis of Signaling Load and Complexity

A reasonable way to analyze the signaling load involved in the information exchange is to count the number of real numbers which are exchanged by each node. In the message-passing algorithm, each node sends a number of $|\mathcal{P}||\mathcal{N}_i|$ real numbers and receives the same amount, per iteration. Let λ_{ave} be the average number of iterations until convergence and assume $L = |\mathcal{N}_i|$

tends to be uniform over nodes for large N in hexagon layout and equals N_{hex}. Then, each node exchanges $\text{I}_{\text{GB}} \approx 2(\lambda_{\text{ave}} L|\mathcal{P}|)$. A greedy technique (e.g. noncooperative game [8]) demands much less information to be exchanged. For instance, each node may exchange only its current own choice (parameter) per iteration, that is, $\text{I}_{\text{G}} \approx 2\lambda_{\text{ave}}$ real numbers. On the other hand, a centralized approach causes a high signaling load over the network. Roughly, each node sends its complete local performance metric and receives its optimal parameter afterwards. That is, each node exchanges $\text{I}_{\text{C}} \approx |\mathcal{P}|^{L+1} + 1$ real numbers.

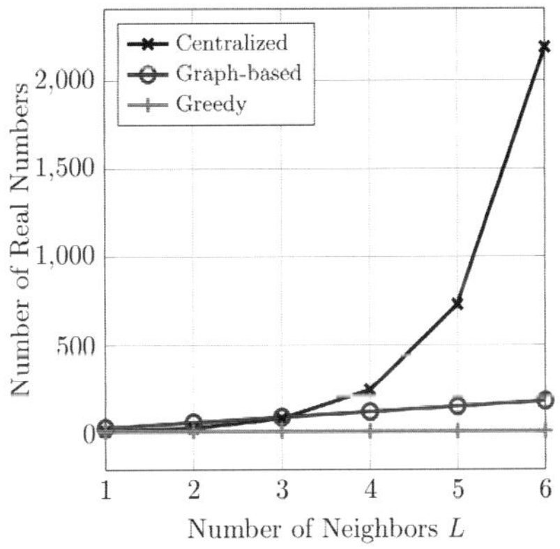

Figure 1. Signaling load analysis showing the amount of information exchange per node against the number of neighbors.

Figure 1 shows the amount of real numbers exchanged per node assuming λ_{ave} equal to five iterations and $1 \leq L \leq N_{\text{hex}} = 6$. Clearly, the function I_{C} increases exponentially and causes more signaling load than the message-passing algorithm for $L > 3$.

4. Simulation Results

The global metric in (1) is investigated in order to evaluate how it behaves statistically in terms of cumulative distribution functions (CDFs) curves. The graph-based technique is compared with the centralized solution and the greedy approach (formulated as a game [8]). Moreover, the 50th CDF percentile of system capacity is evaluated to realize how much gain each distributed technique obtains over the iterations. The MIMO setup of the $N = 7$ communication nodes is such that each transmitter has $N_t = 2$ available transmit antennas, $N_s = 1$ data streams to be transmitted and each receiver has $N_r = 1$ receive antennas, and the codebook $\mathcal{W}$ is as specified in [9]. The signal-to-noise ratio (SNR) equals 20dB and the path loss exponent equals 2 (these simulation parameters and others were taken from [4].)

The graph-based technique appears to reach the optimal solution provided by the centralized approach in terms of system capacity. In Figure 2, the graph-based technique approaches the optimal solution in all the simulation runs, reaching the global optimum in 98% of the cases. That is, the proposed method provides a near-optimal solution. The maximum achievable system capacity is about 5.48 bits per channel use per cell site, reached by both the centralized and the graph-based techniques, while the greedy technique reaches about 4.68 bits per channel use per cell site at the most. As expected, the graph-based technique outperforms the greedy method with a large gain in sum rate. Figure 3 shows the 50th CDF percentile of the curves in Figure 2. The graph-based technique reaches the globally optimum within the first 5 iterations (up to 100 allowed) and the percentage gain obtained over the greedy technique is approximately 33% in system capacity at the fifth iteration. Considering the output at the first iteration, the graph-based approach outperforms the greedy with a percentage gain of about 27% in system capacity. Thus, one may think that the graph-based approach can provide good performance in terms of system capacity even at the very first iteration.

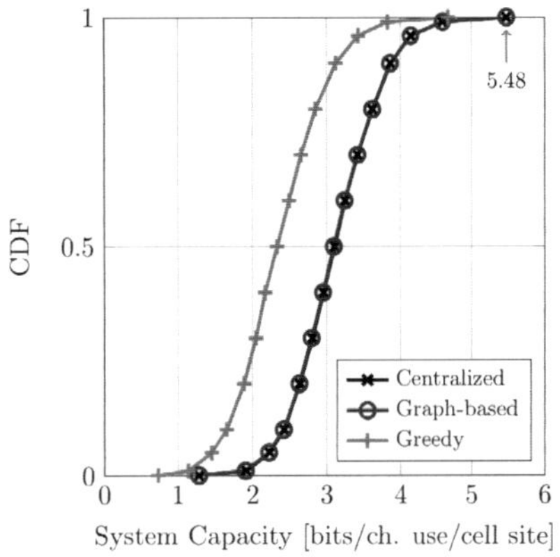

Figure 2. Performance analysis in terms of CDF of system capacity.

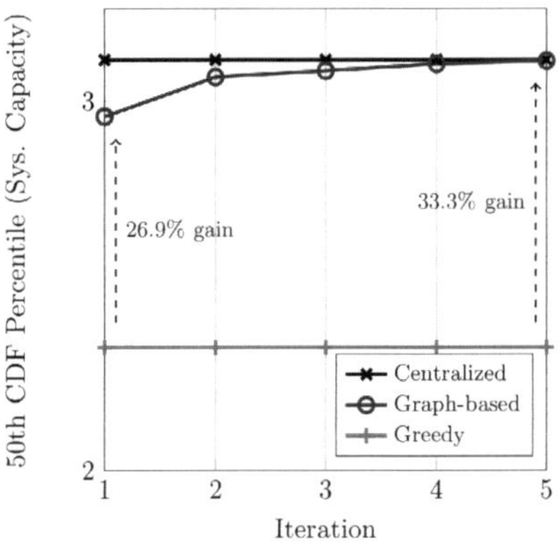

Figure 3. 50th CDF percentile of system capacity against iteration.

5. Conclusions

The graph-based method for distributed parameter coordination considers the impact of nodes decisions on their neighboring nodes. That is, the message exchange is only among neighbors. Such a technique reaches the (near) optimal solution at cost of higher signaling load and complexity compared with the greedy solution, but more efficient than the centralized technique for large network size. As for the numerical results, the graph-based technique provides good gains in system capacity over the greedy solution. It is worthwhile to note that the graph-based approach is totally adaptable to any discrete problem of parameter coordination and any network size. As future studies, one may think of working on message-passing scheduling with faster convergence and message exchange with reduced message size.

Acknowledgments

This work was partially supported by the Innovation Center, Ericsson Telecomunicações S.A., Brazil, and CNPq (306774/2011-9).

References

[1] Han Z and Liu K 2008 *Resource Allocation for Wireless Networks: Basics, Techniques, and Applications* (Cambridge University Press) ISBN 9780521873857

[2] Paulraj, A, Nabar, R and Gore, D 2003 *Introduction to Space-Time Wireless Communications* (Cambridge, UK: Cambridge University Press) ISBN 0521826152

[3] Aji S and McEliece R 2000 *IEEE Transactions on Information Theory* **46** 325 –343 ISSN 0018-9448

[4] Guerreiro I M, Hui D and Cavalcante C C 2013 *EURASIP Journal on Advances in Signal Processing* **2013** 83 ISSN 1687-6180

[5] Telatar, Emre 1999 *European Transactions on Telecommunications* **10** 585–595

[6] Scutari, G, Palomar, D P and Barbarossa, S 2008 *IEEE Transactions on Signal Processing* **56** 1230–1249 ISSN 1053-587X

[7] Kschischang F, Frey B and Loeliger H A 2001 *IEEE Transactions on Information Theory* **47** 498 –519 ISSN 0018-9448

[8] Guerreiro I M and Cavalcante C C 2010 A distributed approach for antenna subset selection in MIMO systems *2010 7th International Symposium on Wireless Communication Systems (ISWCS 2010)* pp 199 –203 ISSN 2154-0217

[9] 3GPP 2012 Evolved universal terrestrial radio access (E-UTRA); physical channels and modulation Tech. Rep. TR 36.211 V11.1.0 Third Generation Partnership Project

IOP Conf. Series: Materials Science and Engineering **67** (2014) 012024 doi:10.1088/1757-899X/67/1/012024

Performance evaluation of power control algorithms in wireless cellular networks

C Temaneh-Nyah, V Iita

Department of Electronics and Computer Engineering, University of Namibia

Email: clementtemaneh@yahoo.com

Abstract. Power control in a mobile communication network intents to control the transmission power levels in such a way that the required quality of service (QoS) for the users is guaranteed with lowest possible transmission powers. Most of the studies of power control algorithms in the literature are based on some kind of simplified assumptions which leads to compromise in the validity of the results when applied in a real environment. In this paper, a CDMA network was simulated. The real environment was accounted for by defining the analysis area and the network base stations and mobile stations are defined by their geographical coordinates, the mobility of the mobile stations is accounted for. The simulation also allowed for a number of network parameters including the network traffic, and the wireless channel models to be modified. Finally, we present the simulation results of a convergence speed based comparative analysis of three uplink power control algorithms.

1. Introduction

Technological advances in the area of wireless communication have given rise to significant increase in the number of mobile wireless devices. The challenge of future wireless systems is to accommodate the increasing demands for wireless services as well as new high-data-rate applications that require performance to meet system Quality of Service (QoS) [1]. An essential radio resource management method in the development of optimal radio transmission like CDMA cellular communication systems is the transmitter power control. Power control intents to control the transmission power levels in such a way that the required QoS for the users is guaranteed with lowest possible transmission powers [2]. Studies also indicate that all forms of electromagnetic radiations adversely affect human and animal health. The biological effects of non-ionizing radiations are extensively examined in a report published by the International Commission on Non-Ionizing Radiation Protection [3]. Transmitting with lowest possible transmission powers in a complex electromagnetic environment can greatly reduce the radiations level and its effect on the operation of radio communication equipments and on biological objects. Mobile radio network based on CDMA technology uses a common frequency band for the organization of traffic channels. Frequency division is used only for duplex. Therefore interference to signals of base stations (BS) are signals of other BS, and interference to signals from the mobile station (MS) are other MS that are in same cell as well as in neighbouring cells. The main problem in the construction of a mathematical model of a CDMA network is to determine the transmission power level in the forward and reverse traffic channels. In a deployed CDMA network, transmission power level of a MS P_{MS} is a function of the signal-to-noise plus interference ratio at the BS receiver input γ_{BS}: $P_{MS} = f(\gamma_{BS})$, and the level of radiated BS partial power P_{BS} is a function of the signal-to-noise plus interference ratio MS receiver input γ_{MS} : $P_{BS} = f(\gamma_{MS})$. The value of the signal-to-noise plus interference ratio γ at a given point in space is dependent on several parameters including the transmitters' power and the propagation loss amongst many others as shown in expression (4) and can be expressed as $\gamma = \varphi(P,..........)$. Thus, obtaining the following expression:

$$P = f(\gamma) = f(\varphi(P,.....)) \tag{1}$$

The solution of (1) is made difficult by its nonlinear nature, as well as for the fact that at any point in space, the calculation of the value of γ requires knowledge of the power emitted by all the network transmitters, but this power cannot be calculated if the value of γ is unknown, which again depends on P. This results in a vicious circle and therefore excludes a simple solution.

Power control is a well-known strategy to minimizing key performance parameters including [4-6]: *Energy usage*: providing the minimum power required for each connection, this implies - reduce battery consumption hence longer battery life expecially for the MS which mitigates frequent battery charging. *Interference Mitigation*: reduces the effects of co-channel and adjacent channel interference. *Improve network capacity:* allows more users to be connected to the network and thus increasing capacity of the network.

Environmental significance: By minimizing the *energy usage* in a wireless cellular communication network through the implementation of efficient PC algorithms, the following environmental benefits can be derived: *Reduce environmental pollution*: lower rate of battery ware out and thrown into the environment. *Human health:* Some studies show that exposure to electromagnetic field has an adverse effect on the human health. The effect depends on the intensity of the electromagnetic field. Therefore achieving a minimum transmit power for each transmitter in the network can significantly reduce the field intensity in the environment and hence less risk to the human being. There has been significant theoretical study and simulation of power control using Signal-to-interference (SIR) based and bit-error-rate (BER) based algorithms. Currently, an algorithm for power control in CDMA network is implemented in Spectrum Engineering Advanced Monte Carlo Analysis Tool (SEAMCAT) and it uses the relationship between power in the traffic channel and bit energy to noise ratio for power control. This data exist only for a narrow range of bit energy to noise ratio, and therefore limits the use of this approach. However, solutions found by this effort may not be applicable in real deployed systems because of simplified assumptions in the system's models such as the use of a rigid structure in defining the fixed distances between the base stations, i.e., the real geographical coordinates of base stations are not considered. Therefore in this paper, we consider implementing the power control by making use of real network data. We consider different power control algorithms in section 2.

2. Power control algorithms

Different algorithms for different wireless communication systems have been studied e.g. centralized, distributed algorithms etc. *Centralized Power Control* involves a central controller with information of all the link gains of the system. Link gains are utilized to find the optimal solution to control the power in all the links simultaneously. However, this algorithm is practically unrealizable due to equipment complexity and bandwidth consideration. In this paper, we consider the *distributed constraint power control* (DCPC) and the fixed step power control (FSPC) algorithms which are examples of iterative algorithms and are considered in [2] and described respectively by the following expressions:

$$\text{FSPC: } P_{MS}(n) = \min\{P_{MS\max},\ P_{MS}(n-1) + \Delta P_{MS} \times sign(e(n-1))\} \tag{2}$$

$$\text{DCPC: } P_{MS}(n) = \min\{P_{MS\max},\ P_{MS}(n-1) + \Delta P_{MS} \times e(n-1)\} \tag{3}$$

$P_{MS}(n)$ is the MS traffic power at the n-th iteration, $P_{MS\max}$ is the maximum transmit traffic power for the MS, ΔP_{MS} is the MS fixed power update step, $e(n-1)$ is the error which is the difference between targeted γ_{BS}^{target} and previous values $\gamma_{BS}(n-1)$ of the signal-to-noise plus interference ratio at the BS, [dB]. The bit energy-to-interference-plus-noise ratio $\gamma_{\mathrm{BS},\, i\, j}$ in the reverse traffic channel between MS and BS with indexes i and j can be estimated by the expression [2]:

$$\gamma_{\mathrm{BS},\, i\, j} = \frac{\left(P_{\mathrm{MS}}^{pc} G_{\mathrm{MS}} G_{\mathrm{BS},\, j}\left(\theta_{i\, j}\right) / L_{i\, j}\right) \times W / R}{P_{\mathrm{N,\, BS}} + P_1 + P_2 + P_{ext}} \tag{4}$$

where

$$P_1 = (1-\beta) \sum_{k \neq i}^{MS,j} v_k \frac{P_{MS,k\, j}^{pc} G_{MS} G_{BS,\, i}\left(\theta_{k\, j}\right)}{L_{k\, j}}, \tag{5}$$

$$P_2 = \sum_{\substack{m=1 \\ m \neq j}}^{N_{BS}} \sum_{l \in \Omega_m} v_1 \frac{P_{MS,\, l\, j}^{pc} G_{MS} G_{BS,\, j}\left(\theta_{l\, j}\right)}{L_{l\, j}} \tag{6}$$

$P_{MS,\, l\, j}^{pc}$ is the emissions power of MS in the reverse traffic channel between MS and BS with indexes i and j respectively; $G_{BS,\, j}\left(\theta_{l\, j}\right)$ is the antenna gain of BS with index j in the direction towards the MS with index i; $L_{k\, j}$ is the propagation loss between MS and BS with indexes k and j respectively; $P_{N,BS}$ is the intrinsic noise power of BS; $\beta < 1$ is the orthogonality factor, which allows us to account for the violation of orthogonality of the spreading pseudo-random sequences; $N_{N,BS}$ is the number of MS served by the BS with index j; Ω_m is the set of MS indexes served by BS with index m; v_k is the mobile user activity which is a random variable with binomial distribution; N_{BS} is the number of BS; W is the chip rate; R is the transmitted information rate.
We can observe that in the case of stationary channels, if $\Delta P_{MS} = 1$, in the algorithm for DCPC then $e(n-1)$ represents the power update for which $e(n) = 0$.
A different algorithm called the Adaptive Step Power Control (ASPC) algorithm can be defined as
$\mathrm{P}(n+1) = \min\left\{\mathrm{P}_{\max}, \mathrm{P}(n) + \Delta \mathrm{P} \tilde{e}(n)\right\}$,
Such that $\tilde{e}(n)$ is an adaptive coefficient for power update express as:

$$\tilde{e}(n) = a(n)\tilde{e}(n-1) + \delta_e u(n) \tag{7}$$

where δ_e is a parameter which determine the rate of power update, $u(n) = sign(e(n)$ is the controlling function. Initial values can be taken as $\tilde{e}(0) = 0$; $u(0) = 1$. Parameter $a(n)$ is defined as follows:

$$a(n) = \frac{1}{2}\left[1 + u(n)u(n-1)\right] \tag{8}$$

i.e.

$$a(n) = \begin{cases} 1, & u(n) = u(n-1) \\ 0, & u(n) \neq u(n-1) \end{cases}$$

Therefore, an Adaptive Step Power Control (ASPC) algorithm can be written as

$$\mathrm{P}(n+1) = \min\left\{\mathrm{P}_{\max}, \mathrm{P}(n) + \Delta \mathrm{P} \tilde{e}(n)\right\} \tag{9}$$

The algorithms FSPC and ASPC has different rate of power update. An example of the power update for these algorithms in the case of $u(n) = 1$ $n = 0, 1, 2, \ldots.$ is shown in table 1 and the plot as shown in figure 1.

IOP Conf. Series: Materials Science and Engineering **67** (2014) 012024 doi:10.1088/1757-899X/67/1/012024

Table 1: Rate of power update

Algorithm	**Power** $P(n)$	**Rate of change**
FSPC	$P(0)+n\Delta P$	linear
ASPC	$P(0)+\Delta P\cdot\delta_e\dfrac{n(n+1)}{2}$	quadratic

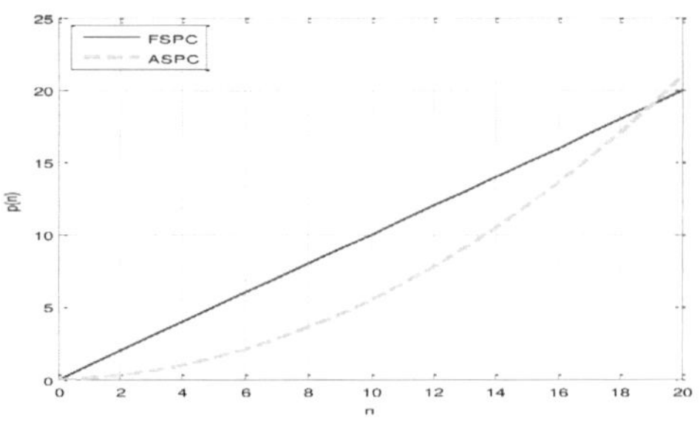

Figure 1. Rate of power update

3. Simulation Results

A simulation platform was developed to simulate and study the power control algorithms (FSPC (2), DSPC (3) and ASPC (9)) for a deployed network while accounting for all the parameters in (4). The relationship between a reverse traffic channel power, the total reverse traffic channel power versus the iteration number for the ASPC, FSPC, and DCPC algorithms for $\Delta P = 2$ $\delta_e = 1$ are presented in figure 2 and figure 3 respectively.

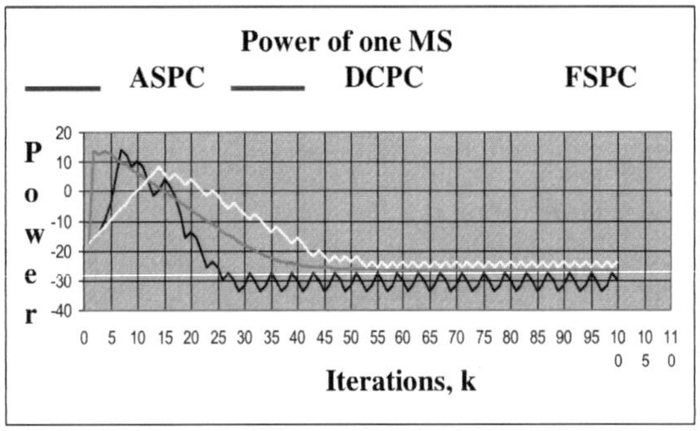

Figure 2. Reverse traffic channel power vs. Iteration number

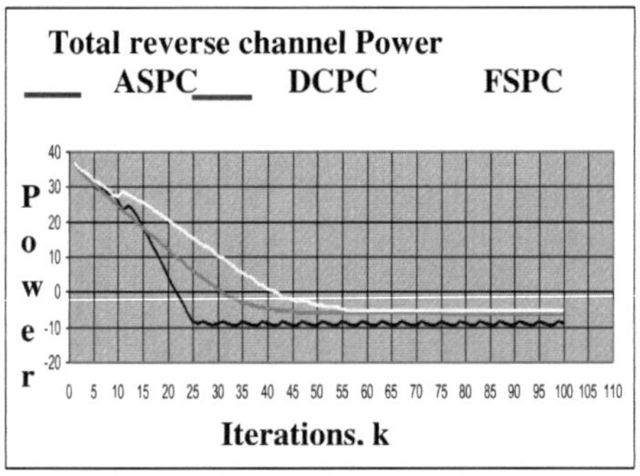

Figure 3. Total reverse traffic channel power vs. Iteration number

4. Conclusion

A platform enabling the simulation of power control in an operational network has been studied. Three power control algorithms have been compared based on the convergence speed. The results shows that the ASPC converges (25 iterations) faster than the DCPC (40 iterations) and FSPC (60 iterations).

5. Reference

[1] Z Zheng 2012 QoS-based Efficient Resource Management in WCDMA Cellular communications, *Journal of Computational Information Systems* **8**: 10, pp. 4249 – 4256

[2] A Dadhich and S Agrawal 2012 Comparative Study of various Power Control Algorithms in WCDMA, *International Journal of Digital Application & Contemporary research* **Volume 1**, Issue **4**

[3] Review of the scientific evidence on dosimetry, biological effects, epidemiological observations, and health consequences concerning exposure to high frequency electromagnetic fields (100 kHz to 300 GHz), **ICNIRP 16/2009**

[4] J. Rulnick and N Bambos 1997 Mobile power management for wireless communication networks *Proc. IEEE INFOCOM,* vol. **3**, pp. 3 - 14

[5] N Bambos and S Kandukuri 2002 Power-Controlled Multiple Access Schemes for Next-Generation Wireless Packet Networks *IEEE Wireless Communications,* pp. 58 - 64

[6] A Dua and N Bambos 2007 Distributed Backlog-Driven Power Control in Wireless Networking *Proc. IEEE Workshop on Local & Metropolitan Area Networks,* pp. 13 – 18.

IOP Conf. Series: Materials Science and Engineering **67** (2014) 012025 doi:10.1088/1757-899X/67/1/012025

Performance analysis of decode and forward cooperative diversity networks with threshold scheme

Shweta Rajoria, Anjana Jain and Prakash D Vyavahare

S. G. S. Institute of Technology and Science, Indore, INDIA

E-mail: shwtrajoria@gmail.com,jain.anjana@gmail.com

Abstract-Cooperative diversity technique is used for achieving high data rate and low BER with low transmit power. The paper proposes a new scheme for selection of first relay having SNR above a predefined threshold level in cooperative diversity network, based on geometric distribution. Closed form expression for outage probability is derived for independent and identically distributed channels with decode and forward protocol. Numerical results show that the proposed scheme not only minimizes resource requirements but also improves the overall outage probability over previously proposed schemes for cooperative diversity network.

1. Introduction

In cooperative diversity networks, cooperation among relay stations help to create spatial diversity in the wireless network even if the individual nodes do not use antenna array for transmission and reception [1], thus improving availability of wireless system. Outage probability analysis for various relaying strategies at high SNR over a Rayleigh fading channel is derived in [2]. The paper proposes, a cooperative network with M relay nodes in which M+1 channels are needed to ensure orthogonal transmission which reduces bandwidth efficiency. To overcome this drawback new cooperative scheme was proposed in [3] that requires C (C< M) relays to forward source information. Further in [4, 5], best relay and N^{th} best relay selection scheme were proposed in order to improve performance and to utilize channel resources efficiently. In best relay selection scheme and N^{th} best relay selection scheme there is excess processing delay at each relay to find best (or N^{th}) best relay. Recently, Al-Tous and Barhumi [6] have proposed a decode and forward cooperative diversity scheme which investigated sharing of two order best relay. This paper proposes a scheme for selection of a relay having SNR above specified threshold level in cooperative diversity network with decode and forward protocol. It is based on geometric distribution to find the first relay having SNR above threshold. The proposed scheme results in reduced processing time and in saving time slots/ channel in the system.

The paper is organized as follows: Section 2 introduces the proposed system model. Section 3 presents the analysis of outage performance of the proposed scheme. Section 4 demonstrates numerical results and section 5 concludes the paper.

2. System model

Source S communicates with destination D directly and with the help of M other cooperating relay nodes via channel with flat Rayleigh fading coefficients as shown in figure 1. The Rayleigh fading channels from source to node and node to destination are assumed to be independent and identically distributed. Consider a Time Division Multiple Access arrangement with M+1 time slots to ensure orthogonal transmission. In the first time slot, the source node broadcasts its signal to the destination node and to the set of M relay nodes R_1, R_2 ... R_M. A minimum SNR threshold level is defined such that any relay node having SNR above threshold has ability to successfully decode the source message.

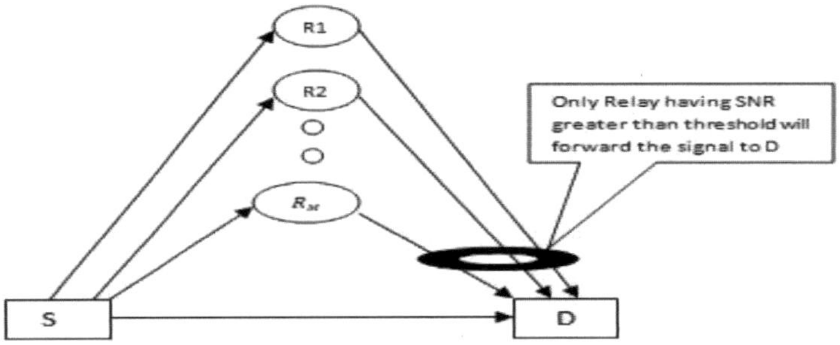

Figure1. Illustration of cooperative diversity network with threshold scheme.

In the second time slot, first relay having SNR above threshold is only allowed to process and forward source information. Here, geometric distribution is used for the selection of first relay having SNR above threshold. Maximum combining ratio technique is used at the destination to combine the direct link signal and the relay signal.

3. Outage probability analysis

Mutual information between source and cooperative node $R_i = 1,2,3,...M$ is given by [4]

$$I_i = \frac{1}{2}\log(1 + \bar{\gamma}\,\hat{x}_i) \tag{1}$$

Here, factor ½ indicates that only two time slots are needed to forward source signal instead of M+1 channels/timeslots as in [3]. $\hat{x}_i = |h_{s,i}|^2$ where, $|h_{s,i}|$ is channel gain between source S and relay station i, defined as a zero-mean circularly symmetric Gaussian random variable. Therefore, $\hat{x}_i$ is exponentially distributed with parameter $\hat{\lambda}_i$ And $\bar{\gamma}$ is the average transmit SNR.

According to the scheme, it is assumed that K^{th} number relay is the first relay having instantaneous SNR above threshold. Mutual information between source and destination using threshold scheme is

$$I_{DF} = \frac{1}{2}\log(1 + \bar{\gamma}(x_0 + \underbrace{x_K}_{K\in 1,2,3..M})) \tag{2}$$

The outage probability P_{out} for required transmission rate, R is given by

$$P_{out} = [I_{DF} \le R] = P\left[(x_0 + \underbrace{x_K}_{K\in 1,2,3..M}) \le {(2^{2R}-1)}\big/{\bar{\gamma}}\right] \tag{3}$$

Where, $x_0 = |h_{s,d}|^2$ and $x_K = |h_{K,d}|^2$ are exponentially distributed with parameter λ_0 and λ_k. As the channel is i.i.d, therefore, $\hat{\lambda}_i = \lambda_0 = \lambda_K = \lambda$.

Let Z be an another random vector, such that, $Z = x_0 + Y$, where $Y = x_1, x_2, x_3, ...x_M$.

Then, equation (3) of outage probability can be rewritten as

$$P_{out} = P\left[Z \leq {(2^{2R}-1)}\big/{\bar{\gamma}}\right] \tag{4}$$

From the above equation it can be visualized that outage probability is CDF of Z at ${(2^{2R}-1)}\big/{\bar{\gamma}}$ as described below.

PDF of Z

PDF of Z is the convolution of the PDF of Y and PDF of x_0. For the Rayleigh fading channel, PDF of x_0 is given by

$$f_{x_0}(x) = \lambda \exp(-\lambda x) \tag{5}$$

By using total probability theorem and having the knowledge of geometric distribution, PDF of Y can be written as

$$f_Y(x) = \sum_{K=1}^{M} P[x_K \mid K^{th} \textit{ relay is first relay above threshold}].P[K^{th} \textit{ relay is first relay above threshold}]$$

$$f_Y(x) = \sum_{K=1}^{M} \lambda \exp(-\lambda x).p.(1-p)^{K-1} \tag{6}$$

Where, p is the probability that randomly selected relay has SNR above threshold value, which is given by

$$p = P[x_k \geq \gamma_{th}] = \exp(-\frac{\lambda \gamma_{th}}{\bar{\gamma}}) \tag{7}$$

After convolving PDF of x_0 and PDF of Y, PDF of Z can be written as:

$$f_Z(x) = \sum_{K=1}^{M} \lambda^2 x \exp(-\lambda x).p.(1-p)^{K-1} \tag{8}$$

and P_{out}, which is CDF of Z, is given by

$$P_{out} = F_Z(x) = \sum_{K=1}^{M} \lambda^2 \exp(-\frac{\lambda \gamma_{th}}{\bar{\gamma}}).\{1-\exp(-\frac{\lambda \gamma_{th}}{\bar{\gamma}})\}^{K-1}.[1-(1+\lambda x)\exp(-\lambda x)] \tag{9}$$

4. Numerical results

In this section numerical evaluation and results for the outage probability of proposed threshold scheme for i.i.d links is being presented. Figure 2 shows results for outage probability as a function of total number of cooperative relay nodes in the system for R = 1/2 bit/sec/Hz, $\gamma_{th} = 1, \lambda = 1$. It may be noted that outage probability performance improves as number of relays in the system increase.

Table 1. Outage probability comparison for M = 2.

Scheme	SNR(5 dB)	SNR(10 dB)
Proposed Scheme	$\mathbf{10^{-3}}$	$\mathbf{10^{-4}}$
$\boldsymbol{N^{th}}$ best relay Selection scheme	$\mathbf{10^{-1}}$	$\mathbf{10^{-2}}$

Furthermore, it is seen that even at the low SNR, proposed scheme gives better results as compared to the previously proposed schemes. Finally, comparison of results of proposed scheme with the results given in [5] is summarized in Table 1. It is seen that for a system with three relays, plot shows negligible outage probability after 15 dB.

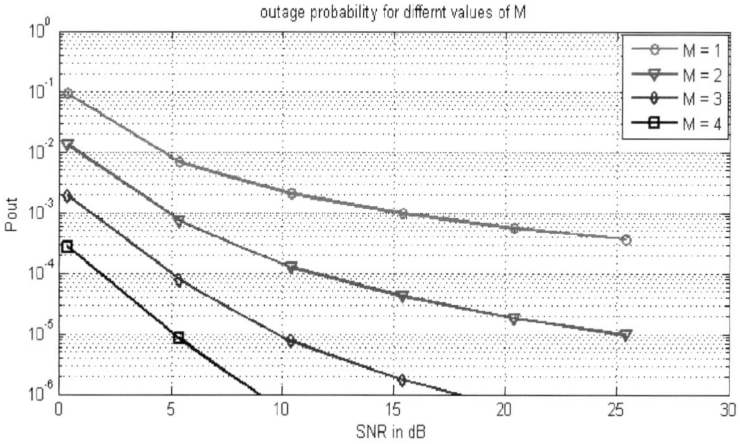

Figure 2. Outage probability performance for R =1 bit/sec/Hz, $\gamma_{th} = 2, \lambda = 1$

5. Conclusion

In this paper, decode and forward cooperative diversity network with threshold scheme over Rayleigh fading channels is being presented and analyzed. Closed form expression for the outage probability is determined for relay links with i.i.d Rayleigh distribution. The main advantage of proposed scheme over the recently presented scheme is that it reduces delay, transmit power and resource. It is observed that for a system with two relay nodes, proposed scheme shows ten percent reduction in the outage probability than earlier proposed scheme at the same SNR of 15 dB.

References

[1] Sendronaris, Erkip E and Aazhang B 2003 User cooperation diversity part 1: system description *IEEE Trans on Comm.* vol 51, pp 1927-38.

[2] Laneman N, Tse D N C and Wornell G W 2004 Cooperative diversity in wireless network: efficient protocol and outage behavior *IEEE Trans on Information Theory* vol 50 pp 3062-80.

[3] Beaulieu C and Hu J 2000 A closed-form expression for the outage probability of decode and forward relaying in dissimilar Rayleigh fading channels *IEEE Comm Lett* vol 10 pp 813-15.

[4] Ikki S S and Ahmed M H 2010 Performance analysis of adaptive decode and forward cooperative diversity networks with best relay selection scheme *IEEE Trans on Comm* vol 58 pp 68-72.

[5] Ikki S S and Ahmed M H 2010 On the performance of cooperative diversity networks with the Nth best relay selection scheme *IEEE Trans on Comm* vol 58 pp 3062- 69.

[6] Hanan Al-Tous and Imad Barhumi 2012 Performance analysis of relay selection in cooperative networks over Rayleigh flat fading vol 58 pp 2- 16.

[7] Simon M K and Alouini M S 2000 *Digital Communication Over Fading Channels* (John Wiley and Sons).

[8] Papoulis A 1991 *Random Variables, and Stochastic Processes* (New York: McGraw-Hill Education).

AUTHOR INDEX